Marx, Engels, and Marxisms

The Marx renaissance is underway on a global scale. Wherever the critique of capitalism re-emerges, there is an intellectual and political demand for new, critical engagements with Marxism. The peer-reviewed series Marx, Engels and Marxisms (edited by Marcello Musto & Terrell Carver, with Babak Amini, Francesca Antonini, Paula Rauhala & Kohei Saito as Assistant Editors) publishes monographs, edited volumes, critical editions, reprints of old texts, as well as translations of books already published in other languages. Our volumes come from a wide range of political perspectives, subject matters, academic disciplines and geographical areas, producing an eclectic and informative collection that appeals to a diverse and international audience. Our main areas of focus include: the oeuvre of Marx and Engels, Marxist authors and traditions of the 19th and 20th centuries, labour and social movements, Marxist analyses of contemporary issues, and reception of Marxism in the world.

Carles Soriano Clemente

Marxism and Earth's Habitability Crisis

From the Metabolic Rift to the Anthropocene

Carles Soriano Clemente
GEO3BCN
Spanish National Research
Council—CSIC
Barcelona, Spain

ISSN 2524-7123   ISSN 2524-7131 (electronic)
Marx, Engels, and Marxisms
ISBN 978-3-031-72536-4   ISBN 978-3-031-72537-1 (eBook)
https://doi.org/10.1007/978-3-031-72537-1

© The Editor(s) (if applicable) and The Author(s), under exclusive license to Springer Nature Switzerland AG 2024

This work is subject to copyright. All rights are solely and exclusively licensed by the Publisher, whether the whole or part of the material is concerned, specifically the rights of translation, reprinting, reuse of illustrations, recitation, broadcasting, reproduction on microfilms or in any other physical way, and transmission or information storage and retrieval, electronic adaptation, computer software, or by similar or dissimilar methodology now known or hereafter developed.

The use of general descriptive names, registered names, trademarks, service marks, etc. in this publication does not imply, even in the absence of a specific statement, that such names are exempt from the relevant protective laws and regulations and therefore free for general use.

The publisher, the authors and the editors are safe to assume that the advice and information in this book are believed to be true and accurate at the date of publication. Neither the publisher nor the authors or the editors give a warranty, expressed or implied, with respect to the material contained herein or for any errors or omissions that may have been made. The publisher remains neutral with regard to jurisdictional claims in published maps and institutional affiliations.

This Palgrave Macmillan imprint is published by the registered company Springer Nature Switzerland AG
The registered company address is: Gewerbestrasse 11, 6330 Cham, Switzerland

If disposing of this product, please recycle the paper.

Contents

1	Introduction	1
2	Earth's Crisis of Habitability and Geosciences	7
	2.1 The Anthropocene. History and Prospective	7
	2.1.1 Empirical Evidence of the Anthropocene	9
	2.1.2 Geologic and Stratigraphic Form of the Anthropocene	14
	2.1.3 The International Chronostratigraphic Chart and the Geologic Time Scale	16
	2.1.4 The Proposal of Formalization by the Anthropocene Working Group	21
	2.1.5 Critique of the Formalization of the Anthropocene in the Geologic Time Scale	27
	References	32
3	Structural Link of the Habitability Crisis to the Capitalist Mode of Social Reproduction	37
	3.1 Marx's Perspective is Relevant for Understanding the Habitability Crisis	39
	3.2 Reproduction and Accumulation of Capital Throughout Commodity Production	41
	3.2.1 The Material Reproduction of Bourgeois Society	42
	3.2.2 Capital as an Automatic Fetish	44

		3.2.3	The Tendency of the Rate of Profit to Fall	47

    3.2.3 The Tendency of the Rate of Profit to Fall — 47
    3.2.4 Social Metabolism with Nature and Metabolic Rift — 50
  3.3 Implications of the Rate of Profit for the Capitalist Systemic Crisis and the Habitability Crisis — 54
    3.3.1 Declining Rate of Profit, Economic Crises, and Capitalist Systemic Crisis — 56
    3.3.2 Revision of the Marxist Debate on the Rate of Profit and the Theory of Crises After the 2007–2008 Financial Crash — 63
    3.3.3 Habitability Crisis, Capitalist Systemic Crisis, and Decreasing Rate of Profit — 76
  References — 79

4 Epistemological Grounds to Transcend the Habitability Crisis — 83
  4.1 The Importance of Philosophy for Scientific Understanding — 84
    4.1.1 Science and Scientific Knowledge — 86
    4.1.2 Natural and Social Science Disciplines — 89
    4.1.3 Natural Laws and Social Production Laws — 91
    4.1.4 Epistemological Paradigms of the Capitalist Society — 94
  4.2 The Anthropocene and Capitalocene as Concepts of the Planetary Crisis — 102
    4.2.1 The Epistemological Perspective of Earth System Science on Nature and the Habitability Crisis — 104
    4.2.2 The Concept of the Planetary Crisis in Capitalocene Narratives — 119
  4.3 Overview of Current Approaches to the Habitability Crisis — 133
    4.3.1 Marxism vs Other Views on the Habitability Crisis — 135
    4.3.2 What Is Marxism? — 140
  References — 145

5 Summary and Conclusions — 149

Index — 157

CHAPTER 1

Introduction

Sometimes it is hard to believe that we are in the midst of a planetary crisis that is seriously threatening life on Earth, leading to the extinction of many known species, and could eventually lead to the extinction of humans themselves. On the one hand, this is because human beings, and especially the bourgeois form of human beings, have developed such a powerful knowledge of reality that gives us a feeling of immortality, a feeling that makes us think no, this cannot happen to us, we can manage and survive whatever terrible future lies ahead of us. On the other hand, waking up on a sunny morning and looking at the transparent waters of the blue ocean or the beauty of a green forest, people are tempted to think, "Is this really happening? Is it not a bit exaggerated? Are we not living a wonderful day in a wonderful world? Such perceptions can lead to skepticism about the habitability crisis on Earth.

Certainly, such perceptions are partly true, but unfortunately, in terms of the planetary crisis of habitability, they are more related to wishful thinking than to a harsh reality that is worsening day by day. Now, the planetary crisis is being directly perceived in everyday life, while the world is being perceived by many as increasingly bleak. It is not simply a question of whether our view of the future is catastrophic or not, but rather of being realistic and making the right choices to overcome a current state of affairs that could make the Earth an uninhabitable planet for many species. It is crucial to recognize that such prospective habitability is not causally inherent in the human condition, in which case the game would already be over and there would be nothing we could do about it. Rather, the current trend toward an inhabitable Earth is causally related to a particular and historical form of human social organization, namely the capitalist or bourgeois form, which implies a particular alienated form of human social metabolism with nature.

As far as we know, life on Earth is an exception in the solar system, and most likely in other planetary systems in the Milky Way. This is simply a fact, an empirical fact, not created by humans, but the result of the natural evolution of matter on Earth. Life on Earth has been threatened several times in Earth's history, and despite the resilience of living things against total extinction, there is no guarantee that it will not happen. Five mass extinction episodes within a relatively short geologic time span are known from strata. For example, about 250 million years ago, 95% of the species known from the geological record were driven to extinction. The rate of extinction today is about two orders of magnitude faster than that of the mass extinction 250 million years ago. It is true that not all species and geological events in Earth's history are recorded in rocks, so there is some uncertainty in extinction estimates, but some uncertainty is the rule for scientific estimates in any field of science. It is also true that humans cannot survive without other living things, so the current rate of extinction is something to be seriously concerned about.

The geopolitical scenario of the ongoing habitability crisis on Earth is that of a slow dismantling of the domination of the world by the Western mode of bourgeois society, a dismantling that started in the past century and is now accelerating. Capitalism expanded from Western Europe to become dominant in the world in a historical process that lasted more than five centuries. The capitalist mode of social production is the last historical form of class society. It consists of the reproduction of capital through the production of commodities, which is achieved through the exploitation of the labor of the working class by the bourgeois class. The reproduction of capital necessarily implies the expansion of the commodity realm worldwide, not only in terms of the activity and life of individuals but also in terms of every form of matter, both inert and living. Thus, the historical process of capitalist expansion meant the proletarianization of every kind of society worldwide and took the historical form of the domination of these societies by the Western bourgeois class. In practice, this was achieved through the Western colonial and pseudo-colonial system and through international institutions dominated by the West, a domination that is now being challenged. Today, the so-called Global South is trying to shake off more than five centuries of Western domination, while the West is in the midst of a profound crisis affecting its moral legitimacy and the material reproduction of Western societies. Western domination of the rest of the world was always based on a sense of supremacy, an exceptionalism of the West with regard to the rest of the world. Amid the Western decadence, such exceptionalism is not even pretended, and so Josep Borrell, the European Union's outspoken foreign policy chief, openly described Europe as an idyllic garden of prosperity and the rest of the world as mostly a jungle that could invade the garden. The crisis of the West and the crisis of the habitability of the Earth are closely related since both are structurally underpinned by the accumulation of capital and the long-term crisis of profitability of capital valorization. In fact, these crises are expressions of the internal contradictions inherent in the system of social production based on the reproduction of capital through the production of commodities. In this regard, it should be recalled that the crisis of habitability will continue on its path regardless of whether capitalism takes a Western or a Global South form.

Fidel Castro used to say that ignorance never helped anyone, and Bertolt Brecht probably gave the best definition of fascism when he said that there is nothing more like a fascist than a frightened bourgeois. That is, fascism is the solution that the bourgeois system pseudo-spontaneously comes up with when the bourgeois status quo can no longer be maintained by the supposedly democratic institutions. Ignorance and fascism are twin brothers that feed on each other, and fascism nests within bourgeois society with the collusion of the bourgeois class. Today, as in the interwar period of the last century, fascism is growing rapidly in the West, and ignorance is also spreading rapidly in Western society, with an obscene manipulation of history by governments, corporate media, and other bourgeois institutions that is an offense to people's intelligence. Today, fascism is already entrenched in many bourgeois institutions, and in addition to ignorance, it relies on fear to germinate and grow like a web within society. In fact, fascism is an expression of the class struggle, it is the unconscious mechanism that the bourgeois society unfolds when it feels threatened. We may be witnessing the prelude to a third world war, driven by a parasitic empire that does not want to lose its hegemony over the world and that is irrationally followed by most countries in the West. All in all, these are symptoms of the decadence of the Western form of capitalist society, which is the original bourgeois form from which capitalism expanded worldwide. If the increasing hostilities turn into an open war, nuclear weapons could be used and the consequences would be dramatic. With all this in mind, one is tempted to think that the current habitability crisis is only of relative importance, and this is true. However, the habitability crisis is already accelerating in the current low-intensity hybrid war, and it will certainly accelerate in the event of a third world war. In any case, the crisis will continue to deepen as long as the capitalist mode remains the dominant mode of social reproduction worldwide.

The purpose of this essay is to provide answers to the questions of why and how human actions have created such a habitability crisis on Earth. It also provides some practical strategies for reversing, not just mitigating, this crisis. The main thesis of this essay is that any practical solution to the habitability crisis must be based on a scientific foundation aimed at establishing a sustainable human adaptation to the planet. This is only possible if both nature and humans are understood within a monistic epistemological paradigm based on dialectics and materialism, in the sense of many philosophers and scientists of the former Soviet Union. The reason for this is that both nature and society are dialectical and material realities, and our thinking about them must reflect as closely as possible such a reality. In other words, our thinking must reflect the ontological reality of the studied objects for a correct practical interaction with them. Otherwise, the interactions between nature and society will not be properly understood at the fundamental level, and a successful practice aimed at overcoming the crisis of habitability could not be undertaken. Today, however, the interaction between nature and society is mostly understood from a dualistic epistemological approach and from an idealist-based monistic epistemology, which prevents a complete scientific understanding of this interaction. The thesis of this book is often developed as a critique of this dualism and of the positivist and idealist understanding of the relationship between society and nature that prevails today.

The first part of the book reviews and discusses the empirical evidence for the planetary crisis and its manifestation in the geologic record, along with the possible formalization of the crisis as a chronostratigraphic unit of the Geologic Time Scale. Although this first part may seem somewhat geologically technical, it is necessary for several reasons. First, the habitability crisis on Earth reveals the deep interconnections between historical human social organizations and our scientific knowledge of the planet. Moreover, these connections claim to transcend the traditional barriers between the social and natural sciences in order to achieve a comprehensive understanding of the crisis. Second, the Geologic Time Scale is the tool by which Earth's history is referred to, and this includes human beings as a product of the evolution of matter on Earth. It is therefore necessary to understand the basis on which Earth system scientists and Anthropocene researchers have approached the planetary crisis in such a tool. Third, the critical analysis of the methodological procedure of natural scientists in formalizing the planetary crisis in the Geologic Time Scale reveals some epistemological flaws in the understanding of the crisis that need to be highlighted.

Most people suspect that the planetary crisis on Earth is somehow related to our current mode of social organization, which is capitalism. However, it is one thing to have a suspicion or an intuition, and it is quite another to be aware of the immanent link between the capitalist mode of social production and the planetary crisis. Such a link must be scientifically demonstrated because this crisis cannot be overcome on the basis of voluntarism and intuition alone, but it must be done on scientific grounds. To this end, the second part of the book examines the socioeconomic roots of the crisis. This requires exposing the foundations of the capitalist system and showing how the planetary crisis is a structural necessity of the reproduction of capital, not just a contingency that can be managed within this mode of production. Since mainstream economics avoids exploring the contradictions of capitalist production and considers failures in the reproduction of capital as mere contingencies rather than structural features, a Marxist approach based on dialectics and materialism is unavoidable. Therefore, Marx's theory of value as exposed in Capital is the reference theoretical framework based on which the ontological link between the crisis of habitability and capitalism is revealed. Thus, this part of the book provides a brief overview of some of the basic principles of capitalist production, especially for those potential readers who are not familiar with Marx's categories in Capital.

The third part of the book critically examines the main epistemological grounds on which the crisis of habitability is approached by the disciplines of the natural and social sciences. It is argued that only Marxism, based on materialism and dialectics, can provide a comprehensive understanding of the planetary crisis and the necessary theoretical guidelines for reversing it if that is still possible. Initiatives to confront the crisis based on the positivist and idealist understanding of the relationship between human beings and nature fail to provide successful practical actions to reverse the crisis for several reasons. First, such a positivist and idealist view understands the contradictions inherent in the reproduction of capital as contingencies that can be formally resolved, rather than as structural features that can only be overcome by overcoming the capitalist system itself. Second, the dualistic epistemological approach to the problem of the relationship between human beings and nature prevents a fully integrated view of this relationship. Third, because of the poor understanding of Marx's theory of value, which is so far the only and best theoretical corpus we have to understand social modes of production in general and the capitalist mode of production in particular.

CHAPTER 2

Earth's Crisis of Habitability and Geosciences

Our understanding of the current crisis of habitability on Earth is possible thanks to the great development of modern geosciences, which now integrate many disciplines of the natural sciences into the broader discipline of Earth system science. This discipline developed in the second half of the twentieth century with the preeminence of global climate studies, which initially faced denial and skepticism. It is now widely accepted that climate is changing on a global scale toward warmer climatic conditions. The development of modern geoscience or Earth system science has been characterized as a Copernican revolution in our understanding of Earth dynamics (Schellnhuber, 1999). Since the late twentieth century, Earth system science and the habitability crisis have been linked to the idea of the Anthropocene and the possibility of characterizing this crisis in the Geologic Time Scale.

2.1 The Anthropocene. History and Prospective

The term *Anthropocene* became popular at the beginning of this century when it was introduced by Nobel laureate Paul Crutzen and his colleague Eugene Stoermer (Crutzen & Stoermer, 2000). The Anthropocene was conceived as a new time in Earth's history, in which human actions are driving changes in Earth's dynamics comparable to those driven by natural forces. Today, the Anthropocene is seen as an ongoing profound transformation of many of the terrestrial geospheres, including the atmosphere, biosphere, hydrosphere, and lithosphere. Research conducted after the introduction of the Anthropocene showed that the perception that human actions are capable of driving significant changes on Earth can be traced back to the nineteenth century. A number of precursors to the Anthropocene proposal—such as Anthropozoic, Anthropogene, Technogene, and others—had been coined earlier to describe the time of Earth's history characterized by such human impacts (Hamilton & Grinevald, 2015; Rull, 2017; Trischler, 2016).[1] The Anthropocene and its precursors are closely related to the unfolding of natural science disciplines—particularly geology—since the eighteenth century, and to the attempt to build a practical tool that can be used to refer to Earth's history based on the stratigraphic record. Today, that tool is the Geologic Time Scale (GTS), a work-in-progress project that undergoes periodic review and improves along with our understanding of Earth's history.

Launching the concept of the Anthropocene within the so-called Earth System science disciplines prompted research on its possible formalization in the GTS. Based on studies of the planetary-scale change of the Anthropocene, geoscientists have asked whether such global change is registered in the stratigraphic record. The affirmative answer to this question has led geologists to consider formalizing the Anthropocene as a new unit of the International Chronostratigraphic Chart (ICC), which serves as the basis for the GTS. In practice, this means that the Earth's history is referred to in a practical tool with a dual nomenclature. On the one hand, the International Chronostratigraphic Chart contains the hierarchical organization of chronostratigraphic units (Systems, Series, Stages) defined on the basis of physical stratigraphic successions. On the other hand, the GTS contains the geochronologic units (Periods, Epochs, Ages) defined on the basis of the units of the ICC adjusted to a linear time scale. In 2009, the International Commission on Stratigraphy (ICS) established an interdisciplinary group of researchers, the Anthropocene Working Group (AWG), with the specific mandate to submit a proposal to formalize the Anthropocene in the ICC and GTS. The AWG submitted the proposal by the end of 2023, and it was rejected by the ICS in March 2024, a rejection later ratified by the International Union of Geological Sciences (IUGS).

---

[1] See Hamilton and Grinevald (2015), Trischler (2016), and Rull (2017) for farther research on the Anthropocene precursors.

Both the formalization of the Anthropocene in the GTS and the proposal of the AWG have generated strong debate from the outset, which has intensified as the date for submission of the proposal by the AWG has approached.[2] For obvious reasons, the refusal to formalize the Anthropocene in the GTS does not end the debate, neither about the formalization of the Anthropocene, nor about the AWG proposal itself, and even less about the Anthropocene beyond formalization.[3] First, because the GTS is an agreement that does not belong to a group of researchers, a commission, or an international organization. Rather, it is a scientifically based world heritage whose origins can be traced back to the seventeenth century (Vai, 2007). Second, because the debate on the Anthropocene has raised some questions about the GTS itself and the rules for formalizing units. Third, because the possibility of a resubmission in the future is not closed.

Here, the empirical evidence supporting the Anthropocene—emphasizing its similarities and differences with comparable changes that occurred during the Earth's history and human history—and the expression of the Anthropocene change in the geologic and stratigraphic record are briefly revisited. This is accompanied by a short exposition of the fundamentals of the GTS, the history, and the current rules to formalize new units, as a basis for undertaking a critical assessment of the problems of formalization of the Anthropocene, and in particular of the proposal by the AWG to formalize the Anthropocene as an Epoch of the Quaternary Period of the GTS.

2.1.1   Empirical Evidence of the Anthropocene

The growing body of empirical evidence that Anthropocene studies have built up over nearly twenty years of research leaves little, if any, doubt about the anthropogenic nature of Earth's transformation. There is only some disagreement about the actual magnitude of planetary change and the different approaches to addressing it. A striking feature of Anthropocene change is the speed with which the various physical, chemical, and biological processes involved are occurring. Empirical data on these

---

[2] For simplicity, I will refer to the Geologic Time Scale (GTS) only.

[3] It took a long time for some units of the GTS to be finally accepted. For example, 30 years of research were needed to finally accept the Holocene (Walker et al., 2015).

processes suggest that they are occurring at accelerated rates, resulting in nonlinear dynamics of the Anthropocene planetary change (Oldfield & Steffen, 2014; Rockström et al., 2009).

For example, the rate of vertebrate extinctions directly caused by human activity exceeds by two orders of magnitude the rate of the major mass extinctions in Earth's history known from the stratigraphic record. The current extinction rate is also about two orders of magnitude higher than the megafauna extinction of the so-called Late Quaternary extinction, which is not considered a mass extinction like the Big Five.[4] Only the Cretaceous-Paleogene (K-Pg) extinction at 66 Ma (million years ago), due to abrupt climate change triggered by a large bolide impact on Earth, shows a higher extinction rate than the current one. However, the role of climate-modifying gases released during Deccan Traps volcanism at about the same age as the bolide impact is still under debate (Sprain et al., 2019). Thus, the rapid extinction rate at the K-Pg boundary may be related to multiple factors rather than a single factor, as is the case with the human-induced extinction rate in the Anthropocene. Biodiversity loss, the homogenization of the world's biota, and the appearance of invasive species due to deliberate or accidental human actions are also occurring at rates unprecedented in human history since the last century. To have an intuition of the potential threat of today's extinctions to Earth's habitability, we need only consider that at the current rate of extinction, a mass extinction of the magnitude of the Big Five will occur in a much shorter period of time. In fact, a recent review of the current extinction worsens previous estimates when demographic decline of population species is considered a symptom of extinction (Finn et al., 2023). Moreover, recoveries from extinction episodes on Earth, including the Big Five mass extinctions, have been shown to have intrinsic speed limits that dampen the amplitude of biodiversity recovery (Kirchner, 2002; Kirchner & Weil, 2000).

---

[4] For comparison of extinction rates between the Big Five mass extinctions and the Anthropocene extinction see Ceballos and Ehrlich (2018) and Soriano (2021). Estimates on extinction rates in these works are based on Barnosky (2011) and Ceballos et al. (2015). For the Late Quaternary Extinction compared to the Anthropocene extinction see Waters et al. (2022).

The Big Five mass extinctions of the geological record are: the Ordovician–Silurian at 444 Ma, the Devonian-Carboniferous at 359 Ma, the Permian–Triassic at 252 Ma, the Triassic-Jurassic at 201 Ma, and the Cretaceous-Paleogene at 66 Ma.

Greenhouse gases, particularly carbon dioxide (CO_2) and methane (CH_4), are being released into the atmosphere by human activities at rates unprecedented in the last 66 million years of Earth's history. The anthropogenic release of CO_2 since the mid-twentieth century is 10 times greater than a well-known release event that occurred 56 Ma ago at the Paleocene-Eocene boundary and 100 times greater than the CO_2 release during the Pleistocene-Holocene transition, which is considered a rapid release in geological terms. Regarding gas emissions during human history, the release of CO_2 during the 6000 years prior to the industrial era was 20 parts per million (ppm), while in the last 200 years CO_2 has increased in the atmosphere up to 100 ppm. Similarly, CH_4 in the atmosphere increased up to 150 parts per billion (ppb) in the 3000 years before the industrial era, and up to 1000 ppb in the last two centuries. The current emission rate of CO_2 is 600 times faster than the pre-industrial emission rate, while CH_4 is being released into the atmosphere at a rate 500 times faster than the pre-industrial rate.[5]

Global surface temperature, loss of tropical forests, desertification, ocean acidification, freshwater pollution, the concentration of solid particles in the atmosphere, and the concentration of plastic particles in the oceans, in short, a daily increasing number of environmental indicators document the anthropogenic impact on the Earth (Steffen et al., 2015a, Goldstein and DelaSalla (eds), 2018). Taken together, the bulk indicators of the planetary crisis show a trend of abrupt increase during the twentieth century, especially since about 1940–1950.

However, when considered individually, a single indicator may show an abrupt increase that is different from that of the bulk indicators. This depends on the physical and chemical properties of the particular parameter, on its role in the Earth's dynamics and its relationship with other parameters, and on the particular human activity related to the parameter under consideration, i.e., industrial production, institutional prohibition, etc. For example, atmospheric CO_2 had already increased abruptly by the mid-nineteenth century, most likely due to the burning of fossil fuels in industrial processes, while atmospheric temperatures increased sharply at the beginning of the twentieth century, mainly due to the greenhouse effect of carbon dioxide. Similarly, the abrupt increase in species extinctions from background extinction levels may have begun in the

---

[5] These estimations are based on Steffen et al. (2016), and Ruddiman et al. (2020).

eighteenth century, or possibly much later, depending on the sensitivity of the different taxa considered (mammals, vertebrates, fish, etc.) to environmental degradation, as well as on the pressure from human activities, which until recently has been greater for terrestrial species than for marine species.

Taken as a whole, the empirical evidence of Anthropocene change suggests that the Earth is now operating in a state that is not analogous to the conditions that prevailed in the Holocene epoch. The empirical indicators of Anthropocene change leave little doubt that the ongoing planetary transformation is limited to the last 300–200 years of human history. Thus, the Earth is now in a state that is also not analogous to any previous period of human history. In fact, the more scientific research on the Anthropocene progresses, the more evidence is gathered that the Earth's transformation since the last 300–200 years has reached an unprecedented scale in human history. In fact, current studies of the Anthropocene locate the *departure in the magnitude* of human impact on Earth within human history in the Great Acceleration of the mid-twentieth century. Given the accelerated rates and the feedback mechanisms between the various processes involved, the planetary change itself is accelerated and increases in magnitude over time. Therefore, the difference between the anthropogenic impact in the Anthropocene and any previous anthropogenic impact is a matter of scale. Put simply, whatever the anthropogenic change on Earth was before the last 300–200 years, it was almost nothing in quantitative terms—magnitude and rate—compared to the Earth transformation in the Anthropocene (Waters et al., 2022; Zalasiewicz et al., 2019).

The evolution of world population is perhaps the best empirical indicator of the difference in scale between the Anthropocene impact on Earth and any previous human impact. Humans are self-producing social beings, and in their metabolic exchange with nature they release a byproduct—which can be seen as an impact—that is now driving a shift in the functioning of the Earth system relative to the Earth's immediate past. An analogy can be drawn with cyanobacteria, whose metabolic activity produced a byproduct, oxygen, which was released through the atmosphere and drove a major change in the functioning of the Earth during the so-called Great Oxidation Episode at about 2.22 Ga (billion years ago). In general, the more individuals—cyanobacteria, humans—the more byproduct is released, so the population is related to the Earth's transformation; but the cyanobacteria analogy ends here. Humans are aware of their own metabolic activity and the transformation that this activity is causing on Earth, whereas cyanobacteria are not. The Earth's atmosphere became oxygenated by a growing population of cyanobacteria with, substantially, the same invariant mode of metabolic exchange with nature. However, the mode of human social metabolism with nature has changed significantly throughout human history, and the number of humans on Earth has evolved accordingly. The evolution of world population is the best empirical indicator of the Anthropocene, because it is the general indicator that synthesizes all other indicators—understood as particular byproducts—measuring the anthropogenic impact on Earth. Therefore, the Anthropocene transformation of the Earth is not only related to the number of human beings but also to the way in which human beings carry out their social metabolism with nature, a metabolism that is historical in the time span of humanity, rather than essentially invariant like that of cyanobacteria.

The evidence is clear: however important human innovations and advances may have been in the past ten thousand years, human population did not increase above 0.5 billion people, whereas from 1700 to 2023 human population increased from 0.6 to 8 billion people, with the sharpest increase since about 1950. Accordingly, the anthropogenic transformation of the Earth during the Anthropocene is occurring on a much larger scale than in the pre-Anthropocene.

Human impact on the Earth is certainly related to the number of humans living on the planet, but the relationship between human population and human impact is not direct and proportional. Rather, it depends on how humans produce, distribute, exchange, and consume their means of subsistence. That is, it depends on how humans are socially and economically organized to reproduce themselves as a collective social being that has a particular interaction or metabolic relationship with the rest of the planet, a metabolic relationship that depends precisely on the kind of social and economic organization. Ten humans are likely to have less impact on the Earth than a thousand humans, and here humans can be replaced by cyanobacteria, beavers, ants, or dinosaurs. However, ten humans organized in one socioeconomic form may have more impact on Earth than a thousand humans organized in another socioeconomic form, and here humans cannot be substituted by any other living or past species. Thus, a comprehensive understanding of the Anthropocene cannot be disentangled from the study of different forms of social organization throughout history. In particular, the current socioeconomic form that has dominated human history in recent centuries must be analyzed from a critical perspective if a scientific understanding of the Anthropocene is to be pursued. Otherwise, the understanding of the Anthropocene becomes a merely descriptive exercise in which historical events—for example, regarding the production and subsequent banning of a certain commodity because of its harmful effects on the environment—are empirically correlated with specific indicators of the Anthropocene. On this basis, no long-term and structural solution to the Anthropocene threat can be proposed, but only specific measures aimed at reducing a given harmful impact.

Unfortunately, such a comprehensive analysis of the relationship between the historical modes of production and their impacts on the Earth has not been undertaken by researchers on the Anthropocene and the Earth system. In fact, the historically determined world population is not even considered as an indicator and a planetary boundary of the Anthropocene transformation on a planet with, by definition, limited resources (Rockström et al., 2009; Steffen et al., 2015b). The correlation between the Anthropocene change and the current mode of social organization—namely, the capitalist mode of social reproduction based on the accumulation of capital through the production of commodities—can be directly inferred on an empirical basis and not from any political, ideological, or moral standpoint. The differential contribution of different countries to the habitability crisis of the Anthropocene, depending on their level of capitalist development, has been documented elsewhere and has been recognized by Anthropocene studies. Ultimately, this is the expression of the differential contribution of different modes of production throughout history, as it reflects the historical process of capitalist expansion on Earth. However, the empirical evidence linking the Anthropocene habitability crisis to the capitalist mode has not been examined in Anthropocene and Earth system science studies to determine whether there is a structural relationship between the two.

2.1.2 Geologic and Stratigraphic Form of the Anthropocene

The imprint of human activity in strata is well known from many scientific disciplines, including archaeology, paleontology, anthropology, biology, geology, and history. Thus, a geologic form of anthropogenic origin is well established, and in terms of the formalization of the Anthropocene, the question is whether the stratigraphic form corresponding to the Anthropocene is qualitatively and quantitatively different from the imprint of human activity in strata in pre-Anthropocene times. The profound landscape transformation of the past 300–200 years has no analog in human history, and given the rate at which this transformation is occurring, it is doubtful that an analog can be found in Earth's history. Most of the Earth's surface changes we see today are the result of human actions over the last century, and this can give us an intuitive approach to quantitatively distinguishing the geologic record of the Anthropocene from that of pre-Anthropocene times. Since the introduction of the term Anthropocene in 2000, research on its possible formalization in the GTS has led to the discovery of an increasing number of stratigraphic markers recorded in an increasing number of paleoenvironmental archives. Such stratigraphic records document the Anthropocene transformation of the Earth system, corresponding to the complex socioeconomic and global organization of capitalism, and most of the stratigraphic proxies discovered have a mid-twentieth century age.In terms of the quantity and variety of key stratigraphic markers and deposits involved, there is no analog in human history. Thus, the human imprint on strata during the Anthropocene has a distinctive signal relative to any previous stratigraphic signal of human activity.

The paleoenvironmental archives that record the imprint of human activity in the Anthropocene are found in a variety of depositional settings, such as speleothems in caves, marine anoxic basins, lakes, ice sheets, coastal marine basins, peat bogs, coral reefs, estuaries, and urban anthropogenic deposits. The most important markers contained in the former archives are isotopes (plutonium, lead, carbon, nitrogen, oxygen), fly ash, plastics, biota (foraminifera, molluscs, ostracods, pollen, zooplankton), and industrial chemical compounds such as polychlorinated biphenyls (PCBs), polycyclic aromatic hydrocarbons (PAHs), and dichlorodiphenyltrichloroethane (DDT).

The suitability of the above stratigraphic markers and depositional facies as the physical references for the Anthropocene unit in the GTS cannot be fully disentangled, neither from the overall structure of the socioeconomic organization that is ultimately responsible for their emplacement, nor from the particular socioeconomic processes underlying the emplacement of a particular marker in a particular paleoenvironmental archive, for several reasons. First, many depositional settings have already been altered to about the same order of magnitude as the Anthropocene transformation, making it increasingly difficult to find pristine environments that record the stratigraphic signal of the Anthropocene impact. Second, most of the proxies considered are directly and indirectly linked to the cycle of capital reproduction based on the production of commodities, to the research on the harmful effects of a particular marker, and to the subsequent social pressure against these undesirable effects. This is equivalent to saying that the occurrence of some of the proxies in depositional environments that can be candidates for the Anthropocene in the GTS is to some extent ultimately related to class struggle and thus express the internal contradictions of capitalist production. For example, the accumulation of lead in bioherms, lake sediments, and ice sheets is in some way related to the history of the production of leaded and unleaded gasoline and the different national and international legislation on the subject; the accumulation of chlorinated pesticides such as DDT in lake sediments and anoxic marine basins has increased worldwide since the 1950s and decreased after its ban; spheroidal carbonaceous particles resulting from the combustion of fossil fuels have been recorded in strata since the nineteenth century, but such stratigraphic records could decrease if a fossil fuel transition occurs. All of these are examples of stratigraphic markers whose particular histories depend on the historical evolution of the cycles of capital accumulation and of national and international legislation aimed at reconciling the interests of capital with the social pressures generated by the harmful effects of capital reproduction.

2.1.3 *The International Chronostratigraphic Chart and the Geologic Time Scale*

The International Chronostratigraphic Chart (ICC) can be understood as a composite stratigraphic succession of the Earth to which the history of the Earth is referred in terms of the processes and events recorded in rocks. An accurate understanding of Earth's history requires that the chronostratigraphic units of the ICC be anchored to as precise a linear time scale as possible. Such a chronometric scale is obtained independently of the ICC chronostratigraphic succession mainly by astronomical tuning of continuous sedimentary strata and by absolute radiometric dating of discrete rocks in the stratigraphic record. The chronostratigraphic succession and the chronometric scale are calibrated using various adjustment techniques to obtain the Geologic Time Scale (GTS), to which geoscientists can refer when studying rocks worldwide. The GTS is intended to be "the tool *par excellence* of the geological trade" (Gradstein, 2012, p. 1 emphasis in the original). Its goal is to provide a standardized tool for geoscientists from many different subdisciplines, and it is therefore eminently practical and pragmatic in nature. Ideally, the GTS should provide a globally correlated, continuous stratigraphic record of the Earth with accurate age estimates, so that orogenic processes, greenhouse episodes, mass extinctions, glaciations, and many other events in Earth's history—such as the human activity in the Anthropocene—can be related to specific, hierarchically organized chronostratigraphic units. Thus, from the very beginning, the ICC and the GTS had the integral multidisciplinary, and holistic approach that is now claimed to be a hallmark of modern Earth system science (Steffen et al., 2016).

The GTS is under continuous revision and evolves roughly in line with our understanding of Earth history. Improvements in dating techniques, analytical geochemical methods, knowledge of depositional systems, fossil evolution, and the geomagnetic field, among many others, have conditioned the configuration of the GTS throughout history. Initially, Stages—the lowest rank hierarchical units of the ICC—were characterized by the approximate chronostratigraphic position and time duration of the stratigraphic section in the type locality of reference. The definition of Stages was usually based on the fossil content and, sometimes, they were merely facies types with local distribution. Global boundary Stratotype Sections and Point (GSSPs) were conceived to overcome the inherent limitations of Stage stratotypes, and to promote global correlation based on continuous stratigraphic records. To this purpose, Stages are now defined on the basis of boundaries with other Stages, and these boundaries must be recognized outside the type locality where they are defined. Besides, GSSPs must be placed within stratigraphic intervals of continuous sedimentation. In this way, GSSPs would allow the reconstruction of a continuous and globally correlated composite stratigraphic section to which the Earth's history could be related. Based on these requirements, an ideal GSSP should be located at a low latitude suitable for cyclostratigraphy and astronomical tuning, should consist of continuously deposited marine cyclic sediments interbedded with volcanic tephra layers suitable for absolute dating, should be fossiliferous and have distinct geochemical signatures and magnetostratigraphy to provide global correlation, and finally, the GSSP horizon should preferably be dated or bracketed between datable horizons (Gradstein & Ogg, 2012).

Chronostratigraphic units defined by GSSPs have, however, their own limitations. In particular, the stratigraphic content of such units is poorly defined, and this may hinder global correlation. Chronostratigraphic units defined by both the GSSP boundaries and the stratotype content, in which the sedimentary record is continuous, multiple correlation markers are identified across unit and unit boundaries, and a chronometric control as accurate as possible is obtained across unit and unit boundaries would be desirable for global correlation. In this way, Earth maps of chronostratigraphic units of different hierarchies—Stage, Epoch, Period—reflecting differences in stratotype content worldwide could be produced along the linear time of the Earth's history. Currently, the chronostratigraphic units of the GTS have a dual time-rock nomenclature (Early-Lower, Age-Stage, Epoch-Series, etc.), which has been reported to be impractical in many cases (Gradstein & Ogg, 2012; Harland et al., 1990; Zalasiewicz et al., 2004). Furthermore, each rock of the stratigraphic record is emplaced within a given time interval and is therefore by definition a time-rock unit. Astronomical adjustment and high-precision cyclostratigraphy of continuous sedimentary strata allow the combination of unit stratotypes and boundary stratotypes, so that chronostratigraphic units can be defined by both their boundaries and their contents, and the dual nomenclature of GTS units can be overcome. This methodology has been successfully applied to several Stages of the Neogene and Paleogene Periods of the Cenozoic Era, although it may pose some problems for much older units of the GTS (Gradstein et al., 2012; Hilgen et al., 2006). Overcoming the dual time-rock nomenclature is one of the main future challenges of the GTS.

The GTS is an agreed-upon convention that aims to provide a standardized tool for relating geologic events recorded in rocks in linear time. The rules for acceptance and revision of units are agreed upon internally within the International Commission on Stratigraphy (ICS) and the International Union of Geological Sciences (IUGS), which are the international organizations in charge of the GTS. It should be remembered that while the GTS is certainly based on a scientific understanding of strata, and more generally of the history of the Earth, the GTS is not science as such, but merely an agreed-upon tool for practical purposes. This means that issues such as the acceptance or modification of units and the rules for formalizing units in the GTS are not only based on scientific grounds, but other political and historical factors intervene in the decisions of the ICS and IUGS. In addition, each chronostratigraphic unit and the GSSP on which it is based have their own peculiarities, and quite often the rules for formalization have been relaxed when accepting new units and GSSP boundaries, or when previous units have been modified according to the particular requirements of the unit under consideration. For example, ice cores containing GSSP boundaries for the Holocene and its internal subdivisions are exceptions in the GTS because most GSSPs are located in rocks.[6] Similarly, isotope fluctuations and Milankovich cycles have only recently been incorporated as primary markers for defining boundaries in GSSPs, and traditionally most boundaries in the GTS have been based on changes in paleontological taxa. Because ice contains valuable information on Holocene climate variability, the ICS has agreed to accept GSSPs in ice cores. This would be an example of a GTS formalization decision driven on a scientific-materialist basis by the specific mode of preservation of Earth's history in ice in the Holocene.

However, other decisions about the rules of the GTS and the formalization of units are based more on political and economic considerations than on scientific understanding. In this regard, it is worth mentioning that the international organizations in charge of the GTS—much like many other organizations worldwide, such as the World Health Organization, the International Monetary Fund, the World Bank, and others—reflect the respective weight of the appointed countries in the global scientific and political scene. Because Western countries in general and Anglo-Saxon countries in particular are dominant worldwide, they are also dominant in these organizations, and thus most of the agreements reached tend to reflect the dominant Western view. This is also the situation for the ICS and the IUGS, which recently froze all activities with the Russian Federation because of the war in Ukraine, but the IUGS did not freeze any activities with the United States because of the war in Iraq, the war in Vietnam, and many other wars in which the United States has been involved.[7] Western countries are also dominant in the ICS and the IUGS, and the decision on the Ukraine war corresponds to their own view on

---

[6] Strictly speaking ice is a rock, but regarding formalization and its preservation potential, it is a different kind of rock than the usual rocks used as archives for GSSP boundaries in the GTS.

[7] This is just one example of the double standards that the collective West applies to the rest of the world, but there are many more. For the statement of the ICS on the Ukraine war see https://stratigraphy.org/.

world geopolitics and reflects their dominant position in these organizations' view. A view that historically has been rather ideologically biased toward non-Western perspectives, especially those of Russian and Soviet-style societies. For example, Soviet society has often been described as ecocidal in its relationship with the environment and nature, despite the fact that the USSR was a pioneer in fields such as ecology and climatology, and in the implementation of environmental protection measures, and had a lower per capita ecological footprint and lower per capita carbon emissions than the United States (Engel-Di Mauro, 2021; Foster, 2015).

Terms such as the Tertiary Period or the Secondary Era have been gradually abandoned in the GTS nomenclature in favor of other terms that more accurately reflect the history of the Earth, especially with respect to the nomenclature of eras, without abandoning the essentially descriptive nature of the GTS nomenclature. However, the Quaternary remains the last period of the GTS, although an alternative term, the *Anthropogene*, was launched in 1922 by the Soviet geologist Alexei Pavlov and was quite popular in the USSR. Notwithstanding the possible technical shortcomings of the Anthropogene as a proposed unit for the GTS, the fact that the Quaternary is a firmly established term in the Anglophone world, and that the official proposal of the Anthropogene was made during the Cold War by Soviet geologists, who had little weight in the international geological organizations, may have contributed to the rejection of the Anthropogene and the subsequent abandonment of any research on its possible formalization.[8]

It is quite common among scientists to believe that research within a particular discipline constitutes *pure science*, and that science can and should be removed from politics, ideology, and any kind of historical and cultural determinations that might structurally condition the development of science. This is only partly true because science is always conducted under structural conditions that usually limit the possibilities of research at any historical time and determine the general path of scientific research. In fact, these historically determined structural conditions regulate what is to be studied and what is not. The Anthropogene case is a good example of such conditions for the particular field of geosciences, but there are others (Heymann & Dalmedico, 2019). Regardless of scientists' individual perceptions of the structural determinants of the process

---

[8] For a brief history of the Anthropogene see Shanster (1973) and Gerasimov (1979).

of scientific research, the truth is that science is not only historically determined, but it is a productive force in all human societies. As Marx noted more than a century ago, in class societies such as capitalism, which is the last and most perfected form of class society, science as a productive force faces human labor as the private property of capital:

> "*Science,* the general intellectual product of social development...appears here as directly incorporated into capital (the application of science as science, separated from the knowledge and skill of the individual worker, to the material process of production), and the general development of society, because it is exploited by capital against labour, because it acts as a productive power of capital over against labour, appears as the *development of capital* ..." (Marx, 1864 *emphasis in the original*).

2.1.4 The Proposal of Formalization by the Anthropocene Working Group

The Anthropocene Working Group (AWG) is a research group of the Subcommission on Quaternary Stratigraphy, which, in turn, participates in the ICS, a subordinate body of the IUGS. The AWG was established in 2009 to assess the potential suitability of the Anthropocene as a unit of the ICC and the GTS. The main conclusion of the AWG arising from the International Geological Congress in 2016 was that the Anthropocene is a stratigraphic reality that started around the mid-twentieth century of which the primary signal would be the plutonium fallout from thermonuclear weapon detonations conducted after the Second World War. On this basis, the AWG prepared a formal proposal that was submitted for its approval to the ICS and the IUGS by the end of 2023. This proposal defines the Anthropocene as the last Quaternary Epoch following the Holocene. After reviewing 12 candidate stratigraphic sections to host the Global Boundary Stratotype Section and Point (GSSP) of the Anthropocene Epoch, the AWG selected a succession of varved sediments in Crawford Lake in Ontario, Canada. In this succession, the primary stratigraphic markers of the Anthropocene would be ^{239}Pu and ^{14}C radionuclides deposited from nuclear weapons testing during the 1940s and 1950s, which would place the starting date for the Anthropocene Epoch in the mid-twentieth century (Waters et al., 2023). The AWG proposal was rejected by the ICS and the IUGS in March 2024.

The formalization of the Anthropocene in the GTS and the proposal of the AWG underwent strong criticisms from the beginning. The Anthropocene has been said to be an issue of pop culture, an academic invention, a political statement, or simply useless (Autin & Holbrook, 2012; Finney & Edwards, 2016; Klein, 2015; Visconti, 2014). However, as AWG members have already acknowledged, criticism is part of the normal scientific discussion, and critiques of the Anthropocene have helped to refine the final AWG proposal. These critiques have also helped to highlight some shortcomings of the ICC and GTS with respect to the formalization of units (Luciano, 2022; Soriano, 2024). Overall, the AWG has responded to criticism in a detailed, rigorous, scientific, and geologically sound manner. Issues such as the preservation potential of Anthropocene series, the global correlation potential, and the short time span encompassed if a mid-twentieth century base for the Anthropocene is accepted, and the question of how chronostratigraphic units relate to major Earth changes have been addressed by the AWG in detail (Zalasiewicz et al., 2017).

Some critics of the Anthropocene formalization seem to have misunderstood the instrumental and practical nature of the GTS. It is a tool to facilitate correlation, primarily among geologists, but not only, as scientists from other natural science disciplines are increasingly involved in modern Earth system science. With the advent of the Anthropocene, the human and social sciences must also be involved. The construction of this tool is certainly based on a scientific understanding of the history of nature, including human history, but also on practical purposes that vary according to the scientific disciplines involved, their goals, their techniques, and other aspects. Such a tool cannot be confused with the scientific understanding of Earth history itself, nor can the rules of formalization of the ICC/GTS be confused with being merely scientific. For these reasons, an agreement is needed, but any agreement is by definition not scientific, but political, regardless of the scientific grounds on which it is based. Moreover, it should not need to be reminded that any decision of a committee on formalization is based on the subjective perception of individuals about practicality, compliance with rules, etc., which is also a political issue.[9]

---

[9] In fact, some decisions of the ICS and IUGS regarding the acceptance of chronostratigraphic units may appear somewhat arbitrary, as in the case of the Tertiary and Quaternary (Knox et al., 2012).

Among the concerns raised by critics of the Anthropocene are that the socioeconomic features implicit in its determination are not a geological matter, that the concept of the Anthropocene comes from outside stratigraphy, and that the supposed inductive character of geological sciences and sciences, in general, is violated in favor of a deductive analysis that would not be characteristic of geology. Although the GTS has been agreed upon primarily by earth scientists, it must be remembered that this instrument is not the private property of the earth sciences, and that scientific disciplines do not operate in isolation from each other and from the socio-historical determinations in which they evolve. Insofar as the transformation of the Earth and its manifestation in strata is driven by human actions carried out under the laws of capital reproduction, a comprehensive understanding of the Anthropocene explicitly requires that socioeconomic processes in general, and the socioeconomic processes of the capitalist mode of production in particular, be integrated with natural processes if a thorough understanding of the subject is to be pursued. Otherwise, the understanding of the Anthropocene will be mutilated and the practical measures to confront this great threat will fail.

It has also been argued that stratigraphy, and thus the GTS, deals with the geological past, whereas the AWG proposal focuses on the present and depends on future scenarios (Autin & Holbrook, 2012; Finney & Edwards, 2016; Rull, 2018). However, a mid-twentieth century age for the base of the Anthropocene documents about 70 years of stratigraphic record, and "... the case being made for the Anthropocene rest solely on evidence documented within *existing* strata that represent *past* events, as it must" (Zalasiewicz et al., 2017). Moreover, the current trend in most scientific disciplines is the ability to predict future scenarios, and stratigraphy is no exception. Weather forecasting, volcanic eruption prediction, depositional environment simulation, financial market prediction, and many other examples from both the natural and social sciences show the increasing predictive nature of science. Predictions can fail, and some disciplines make more reliable predictions than others because of their internal consistency and scientific development. But overall, they are not speculations but are based on scientific understanding of the objects studied. Thus, the evolution of the human population can be predicted for the near future based on past and present data and provided that some boundary conditions are maintained, namely the capitalist mode of production based on the reproduction of capital; the trajectory of the

Earth system can also be predicted with today's understanding of Earth history and with assumptions on key parameters (Steffen et al., 2018).

The inductive versus deductive dichotomy of science, and thus of geology, is a false debate based on some epistemological misunderstandings. The traditional dualism of induction and deduction has long been transcended by a dialectical and materialist view in which both are seen as inseparable moments of any scientific understanding of reality.[10] In short, inductive analysis based on empirical facts alone is impossible in the absence of conceptualizations obtained by deduction that form a theoretical framework on which induction is based, and deductive analysis is equally impossible in the absence of induction based on empirical evidence. When a geologist collects samples from a dike in order to obtain the Anisotropy of Magnetic Susceptibility (AMS) to infer magma flow, such a geologist is relying on a theoretical background that determines the sampling strategy to obtain the expected results, which are then subject to analytical induction. When the same geologist samples a sedimentary succession to obtain the magnetostratigraphic profile, the sampling strategy is different from that required for AMS studies, and it is determined by the theoretical concepts of magnetostratigraphy and sedimentology. When Earth scientists sample paleoenvironmental archives in search of possible candidates for a GSSP for the Anthropocene, their goals and expected results are determined by the theoretical background of Earth system science, geology, stratigraphy, and a minimal understanding of the productive processes of modern society. For example, these geoscientists will not be looking for DDT signatures in a stratigraphic horizon from the nineteenth century or the Neolithic. These examples illustrate how scientific knowledge proceeds: a deductive conceptual background is needed to obtain empirical data, which in turn is used to modify the theoretical background by inductive reasoning. In fact, these examples illustrate the teleological character of scientific research understood as intellectual labor, a teleological character that is also immanent in the GTS, as shown by their iterative modifications throughout history based on deductive-inductive rationale.

A major criticism of the AWG proposal is based on the diachronic nature of the human imprint on strata. Rather than a formal unit of the GTS, it is argued that the Anthropocene should be considered an informal

---

[10] For an extended development of the role of induction an deduction in logic and in theoretical understanding see Rosental (1962) and Ilyenkov (1982).

event in Earth's history, similar to other diachronic events such as the Great Ordovician Biodiversification Event or the Great Oxidation Event, whose diachronism does not allow for a nearly synchronous and globally correlated starting point (Gibbard et al., 2022; Walker et al., 2024). The human imprint on the environment and its geologic and stratigraphic form are certainly diachronic and began roughly coincident with the appearance of humans on Earth. However, these imprints were not quantitatively significant until humans were organized in a social mode of production that is qualitatively different from any other social mode of production in human history. This mode is capitalism, the first mode of production that is global and has allowed an unprecedented development of the productive forces, including science, but which also implies an unprecedented impact on the Earth due to its particularities. The development of capitalism on Earth is also a diachronic, historical and still ongoing process, but only at a certain degree of this process a threshold is crossed and the human impact on the environment and its stratigraphic form are both quantitatively significant in relation to previous forms. The proposal of the AWG has addressed precisely the quantitative departure of the human impact on the Earth and its stratigraphic form. In other words, it has addressed a change of scale in the relationship between humanity and nature, which is based on the particular social metabolism with nature based on of the reproduction of capital, although the AWG does not explicitly mention this. In proposing ^{239}Pu and ^{14}C radionuclides in sediments of Crawford Lake in Ontario as the primary stratigraphic marker of the lower boundary of the Anthropocene, the AWG is simply following the GSSP requirements of global correlation and synchronicity.[11]

The shift in the humans' impact on Earth and of its expression in strata shown by the Anthropocene is both a quantitative and qualitative change that is based on a change in the qualitative relationship between humans and nature. This can be best understood with Hegel's law of dialectics of reciprocal transformation between quantitative and qualitative changes in the general motion of matter, and with an understanding of the relationship between phenomena and essence as a dialectical unit in which observed phenomena appear shaped by multiple mediations of underlying

---

[11] For an extended unfolding of the debate between the Anthropocene as an informal event of the GTS and the AWG proposal, including the importance of philosophy regarding formalization, see Soriano (2024).

essence, which may include the appearance of phenomena as opposite of essence (Saoserov, 1960).

Thus, on the one hand, the quantitative change in human impact in the Anthropocene is based on the qualitative specificity of the capitalist mode of social reproduction with respect to earlier modes of social production in human history, which did not produce such a quantitative leap. Understood in this way, the quantitative change of the Anthropocene marks a turning point in the relationship between humans and nature, without comparison to previous forms of this relationship. A turning point that has its origin and subsequent development in the particular type of social metabolism with nature that is governed by the reproduction of capital rather than by human beings themselves (Foster, 2022).

On the other hand, the essential and structural *causa finalis* of the Anthropocene major shift in Earth dynamics is the contradiction inherent in the reproduction of capital as governed by its constitutive laws. This essential contradiction is expressed in environmental indicators and in strata, which are its phenomenal expressions mediated by the laws or principles of biological processes, the physics and chemistry of the atmosphere, and the principles of geology, such as Steno's law of superposition of strata. Moreover, the dynamics and history of capitalist production, including the class struggle inherent in this mode of production, also mediate the phenomenal expression of the essential capitalist contradiction. For example, social pressure against harmful effects on human health and environment may force the abandonment of the production of a given commodity, and hence modify its stratal expression. All these mediations explain the diachronicity of the phenomenal expressions, both environmental and stratigraphic, and of the Anthropocene shift within the history of capitalist production, even though the bulk of indicators show a major break around the mid-twentieth century. The relationship between the essential roots of the Anthropocene crisis and its phenomenal forms can be illustrated by crude analogies with examples in nature. For example, the Sun seems to move around the Earth, but scientific research has shown that reality is just the opposite. An earthquake is the phenomenal expression of plate boundary friction, which is the essential determination underlying the phenomena, and it is mediated, for example, by local fluid pressure, which may result in earthquakes of different magnitudes, fault ruptures, etc.

2.1.5 Critique of the Formalization of the Anthropocene in the Geologic Time Scale

Studies of the Anthropocene are closely related to the development of modern Earth system science, a discipline that has opened up the possibility of consciously acting within the Earth system on a global scale and on a scientific basis in order to keep the planet habitable. The Anthropocene introduces a new methodological approach to our understanding of Earth's history and in the formalization of units in the GTS. In terms of formalization, prior to the Anthropocene, units of the GTS were typically formalized once the stratigraphic content from which the Earth's history is inferred was reasonably well known. From an epistemological perspective, this means that knowledge progressed from data observed in local strata to the inference of global phenomena. In the Anthropocene, knowledge proceeds in the opposite direction: from observed global phenomena to their expression in local strata. Thus, in the Anthropocene, the human-induced crisis of habitability was known not from strata but from global environmental indicators and then sought in strata. The logic of this approach is that if we know from the geologic record of previous planetary crises similar to the current one, it is reasonable to hypothesize that a stratigraphic signature of the ongoing planetary crisis can be found. Although the search for empirical evidence as a method for validating theories is common in science, this procedure poses some problems in terms of formalization in the GTS, because historically events in Earth's history have been inferred from the geological record, and so the formalization in the GTS reflects this methodological procedure.

In general, the names given to chronostratigraphic units in the GTS nomenclature try to stay as close as possible to mere descriptions, usually avoiding conceptualizations about the history of the Earth. The rationale behind this option is based on the instrumental and practical character of the GTS, that is, on the fact that the GTS is an agreed-upon correlation tool to which the Earth history is referred, but it is not the Earth history as such. Hence, the content of names of most chronostratigraphic units is usually the etymological content of the term. For example, most names of Eons and Eras describe the chronological presence of life in strata: Paleozoic (old life), Mesozoic (intermediate life), and Cenozoic (new or recent life); Phanerozoic (visible life), Proterozoic (earlier life). As recommended by the ICS, Ages—the lowest hierarchical chronostratigraphic units of the GTS—combine their names with the names of geographic features near the stratotype area of the GSSP and "-ian" or "-an" endings. This is also an ICS recommendation for Epochs, although it has not been followed for most Epochs of the Cenozoic Era. For example, Holocene stands for "entirely new or recent," Pleistocene for "newest," Pliocene for "newer," Miocene for "less new," Oligocene for "few new," Eocene for "dawn new" and Paleocene for "old Eocene." These names ultimately refer to the successive appearance of life in strata following the Cretaceous-Paleogene mass extinction at about 66 million years ago, and they were coined by geologists of the nineteenth century.

Following this Cenozoic tradition and the descriptive methodology of the GTS, the AWG has proposed the name Anthropocene—etymologically "new or recent Anthropos or human"—for the new Epoch of the GTS. The first problem with the name Anthropocene is that there is not any "old or less new Anthropos" in the chronostratigraphic units of the GTS, although the stratigraphic record of the human activity before the proposed basal boundary of the Anthropocene is well known. It could be argued that a chronostratigraphic unit corresponding to a "less new Anthropos" is lacking because stratigraphic sections that meet the GSSP requirements for formalization are also lacking, and that the Anthropocene could be accepted as an epoch in the meantime while research on candidate sections that host a GSSP for, say, a Paleoanthropocene unit is successful. However, this option would probably imply a reorganization of the Holocene, Quaternary, and other chronostratigraphic units of the Cenozoic, and the decision-making bodies of the ICS and IUGS have traditionally been reluctant to make drastic changes to the GTS. Such a reorganization is obviously outside the strict competence of the AWG, since it would involve many of the upper units of the GTS. Nevertheless, it is probably time to have consistent nomenclature guidelines throughout the GTS, since currently only Ages, Eons, and Eras have such consistent nomenclature, but not Epochs and Periods. In general, the rules and criteria necessary to formalize and name chronostratigraphic units should be as consistent as possible throughout the GTS (Luciano, 2022). Claims that the names of some Cenozoic units are already well established and have a long tradition are not really arguments, at least not scientific arguments. Blacks have been slaves in the Americas for several centuries, so it was once a well-established tradition, but they are no longer slaves.

The AWG could have followed the ICS recommendation to give to Epochs of the GTS names of geographic origin preferably ending with "ian" or "an" suffixes, as is the case for Guadalupian, Terrenuvian, and others.[12] For the sake of consistency throughout the GTS nomenclature, this option would have probably implied a revision of many names of the lower rank chronostratigraphic units, not only of the Cenozoic, but also of the Mesozoic and Paleozoic Eras. Thus, the AWG could have chosen to combine the name of geographic features in the vicinity of the area of the selected candidate site for the GSSP of the Anthropocene with "ian" or "an" endings. In this way, a possible name for the major Anthropocene change on Earth as recorded in strata could have been Ontarian (Soriano, 2024). However, the AWG borrowed, without proper critical revision, an already established term with a conceptual content that goes beyond chronostratigraphy and GTS, a conceptual term that concerns the historical relationship between humans and the planet. Earth system and Anthropocene researchers have explicitly identified the current planetary crisis to the activity of modern humans on Earth. In particular, these researchers have identified the mid-twentieth century as the turning point of the human impact on Earth in quantitative terms, and this has left clear stratigraphic signals. The Great acceleration charts widely used by Anthropocene scientists are clear empirical evidence of this turning point, and for this reason, pinpointing the onset of the planetary crisis and its stratigraphic manifestation by the mid-twentieth century is roughly correct (Steffen et al., 2015a). However, the naming of the planetary crisis and its chronostratigraphic form in the GTS as the Anthropocene is, at best, ambiguous with regard to the historical relationship between humans and nature and its related impact on Earth. The reason for such ambiguity is that this "new or recent Anthropos or human" whose activity underlies the habitability crisis on Earth is nothing other than the globalized capitalist Anthropos. That is, it is the concrete Anthropos whose social reproduction consists in the accumulation and reproduction of capital by means of commodity production raised to a global level. This is not properly reflected by the ambiguous "new or recent Anthropos" in the term Anthropocene, and the reason is that the concept underlying this term is also ambiguous.

Given the ambiguity of the Anthropocene concept since the term was coined and the uncritical adoption of this term by the AWG, it is understandable that other terms corresponding to less ambiguous conceptualizations of the planetary crisis have been proposed for the

---

[12] For the guidelines about GTS nomenclature, see ICS website https://stratigraphy.org/guide/chron.

GTS. For example, Capitalocene has been proposed to substitute Anthropocene as the name of the new Epoch in the GTS, and this term has undergone subsequent conceptualizations in the humanities and social sciences (Malm, 2016; Moore, 2017, 2018). Capitalian or Capitalinian have been proposed to name the lower Age of the Anthropocene Epoch in the GTS (Foster & Clark, 2021; Soriano, 2020). To be clear, none of these proposals make much sense with the current configuration and nomenclature of the GTS. The Capitalocene has similar problems to the Anthropocene, because it stands for "new capital" while an "old capital" is missing in the GTS. Besides, although the dominant forms of capital have varied throughout history, this is only a formal change, for the essence mode of the social reproduction and of capital itself has remained the same. As for Capitalian or Capitalinian, these terms would break today's consistent nomenclature of the lowest rank chronostratigraphic units of the GTS, which derive their names from the names of geographical features near the stratotype area of the GSSP. In any case, Capitalocene and Capitalian or Capitalinian would imply a profound conceptual modification of the GTS, and not only more or less formal modifications concerning names, because the different modes of production in human history would be reflected for the first time in the GTS. Nevertheless, this is a direct implication of the Anthropocene Epoch itself, where some concrete way of understanding the relationship between geological history and social history specific to the Anthropocene Epoch is implicit, and its ambiguity can only be precise at the level of the geochronological Ages of the GTS.

In terms of nomenclature, a possible solution for the Anthropocene in the GTS might be to replace the *Quaternary* Period with Pavlov's *Anthropogene* Period based on the first fossil appearance and subsequent evolution of the genus *Homo* on Earth. However, further research on the appearance of the genus *Homo* in strata would be probably needed to discard diachronicity. These criteria could serve as a basis to reformulate the Anthropozoic of the Italian geologist Antonio Stoppani and to define it as an Era following the Cenozoic and starting at the base of the Gelasian Age (Rull, 2020). In this way, the specificity of the ongoing planetary transformation with respect to any previous similar transformation in the Earth's history and in human history would be reflected in the GTS, while the descriptive character of the GTS would be respected by recording the appearance of humans in strata and their subsequent historical evolution as a "new Anthropos."

The main flaw of the Anthropocene, however, lies beyond strata, but at the same time, it has implications for stratigraphy and the GTS. As clearly acknowledged by Anthropocene scholars, the Anthropocene is a concept that emerged from the Earth system science community, not from stratigraphy. Notwithstanding the Anthropocene has been vaguely correlated with modern society, the Industrial Revolution, and later on the Great Acceleration of the mid-twentieth century, this correlation has not been explored in the Earth system studies in terms of what is the fundamental cause underlying the Anthropocene as a major shift in the Earth's functioning since the Holocene. Therefore, the term Anthropocene implicitly identified from the beginning, and perhaps inadvertently for Earth system scientists, this major shift to a general Anthropos rather than to a particular and historical form of Anthropos. As a result, the Anthropocene shift is not fully understood on a scientific basis and, more importantly, the crisis of Earth habitability that the Anthropocene has brought to the fore is not confronted on a scientific basis. Due to the lack of research on the deep roots of this Earth shift, research on the Anthropocene has been carried out only at a phenomenological level, rather than trying to understand the essential economic roots of the crisis and how they are linked to the observed phenomena. This has led to a dichotomous understanding of the Anthropocene. On the one hand, as a purely technical issue related to the GTS and the dynamics of the Earth system, which has led to the conception of a "scientific" Anthropocene placed at the "analytical level" and allowing a "very precise, strict understanding." On the other hand, as a less precise issue that "begets criticism and debate," corresponding to the Anthropocene in the humanities and social sciences, and is placed at the non-scientific "consequential metalevel" (see Fig. 3 in Zalasiewicz et al., 2021).

In the end, the dichotomy about the Anthropocene conducted by Anthropocene and Earth system scientists is nothing else than a new form of the old dualism between the natural sciences and social sciences, which in turn is a form of the cardinal problem of philosophy: the relationship between thinking and being (Engels, 1946). Such a dichotomy must be transcended if the habitability crisis of the Anthropocene is to be overcome, but it is impossible to be transcended with the current epistemological approach of the Anthropocene studies (Soriano, 2022). First, because the scientific character of the social sciences is negated, as if social history could not be scientifically understood on a materialist and dialectic basis, as is natural history regardless of whether scientists of the natural sciences are aware or not of the epistemological approach they follow in their research. Second, because the relationships between the various subfields of the social sciences and natural sciences, which are commonly represented as arrows in the so-called "integrative and extended multilevel Anthropocene concept" diagram and in the Bretherton diagram, are merely formal (Steffen et al., 2020; Zalasiewicz et al., 2021). That is, these relationships are devoid of any essential causal concatenations other than those directly observed at the phenomenological level. Third, and most importantly, in terms of the planetary crisis, Anthropocene scholars understand the relationship between the social and the natural upside down. Namely, the social is not at the "consequential metalevel" as claimed by Anthropocene researchers, but just the opposite. Being synthetic, the habitability crisis of the Anthropocene is ultimately underpinned by the causal mechanisms of the accumulation of capital at a declining rate of profit, and for this reason, the crisis is an expression of the fundamental contradiction inherent to the capitalist mode of social reproduction. Although it is true that the planetary crisis has consequences in economics, politics, philosophy, etc., this is a mere phenomenological description of ongoing facts, while the lack of research aimed at investigating the ultimate determinants of the crisis leads to its misunderstanding. In summary, the essential socioeconomic roots underlying the Anthropocene crisis must certainly be distinguished from their phenomenal expression in the environment, geology, and stratigraphy with the particular categories of the economy. However, the dualistic approach of Anthropocene researchers does not permit tracing the concrete causal concatenations linking the fundamentals of the crisis and its phenomenal expressions, which would allow for a truly comprehensive and integrative understanding of the planetary crisis, something that the crisis itself demands.

REFERENCES

Autin, W., & Holbrook, J. (2012). Reply to Jan Zalasiewicz et al. on response to Autin and Holbrock is the Anthropocene an issue of stratigraphy or pop culture? *GSA Today, 22*, e23.

Barnosky, A. D., Matzke, N., Tomiya, S., et al. (2011). Has the Earth's sixth mass extinction already arrived? *Nature, 471*, 51–57.

Ceballos, G., & Ehrlich, P. R. (2015). Accelerated modern human-induced species losses: Entering the sixth mass extinction. *Science Advances, 1*, e1400253.
Ceballos, G., & Ehrlich, P. R. (2018). The misunderstood sixth mass extinction. *Science, 360*, 1080–1081.
Crutzen, P., & Stoermer, E. (2000). The Anthropocene. *Global Change Newsletter, 41*, 17–18.
Engel-Di Mauro, S. (2021). *Socialist states and the environment*. Pluto Press.
Engels, F. (1946). *Ludwig Feuerbach and the end of classical German Philosophy*. Progress Publishers.
Finn, C., Grattarola, F., & Pincheira-Donoso, D. (2023). More losers than winners: Investigating Anthropocene defaunation through the diversity of population trends. *Biological Reviews*. https://doi.org/10.1111/brv.12974
Finney, S., & Edwards, L. (2016). The "Anthropocene" Epoch: Scientific decision or political statement? *GSA Today, 26*, 4–10.
Foster, J. B. (2015). Late soviet ecology and planetary crisis. *Monthly Review, 67*, 1–20.
Foster, J. B. (2022). *Capitalism in the Anthropocene*. Monthly Review Press.
Foster, J. B., & Clark, B. (2021). The Capitalinian: The first geological age of the Anthropocene. *Monthly Review, 73*, 1–16.
Gerasimov, I. P. (1979). Anthropogene and its major problem. *Boreas, 8*, 23–30.
Gibbard, P. L., Bauer, A. M., Edgeworth, M., et al. (2022). A practical solution: The Anthropocene is a geological event, not a formal epoch. *Episodes, 45*, 349–357.
Goldstein, M., & DelaSalla, D. (Eds.). (2018). *Encyclopedia of the Anthropocene*. Elsevier.
Gradstein, F., & Ogg, J. (2012). The chronostratigraphic scale. In F. Gradstein, J. Ogg, M. D. Schmitz, & G. M. Ogg (Eds.), *The Geologic Time Scale 2012*. Elsevier.
Gradstein, F. (2012). Introduction. In F. Gradstein, J. Ogg, M. D. Schmitz, & G. M. Ogg (Eds.), *The Geologic Time Scale 2012*. Elsevier.
Gradstein, F., Ogg, J., & Hilgen, F. (2012). On the Geologic Time Scale. *Newsletter on Stratigraphy, 45*, 171–188.
Hamilton, C., & Grinevald, J. (2015). Was the Anthropocene anticipated? *The Anthropocene Review, 2*, 59–72.
Harland, W. B., Armonstrong, R. L., Cox, A. V., et al. (1990). *A Geologic Time Scale 1989*. University Press.
Heymann, M., & Dalmedico, A. D. (2019). Epistemology and politics in the earth system modeling. *Journal of Advances in Modeling Earth Systems, 11*, 1139–1152.
Hilgen, F., Brinkhuis, H., & Zachariasse, J. (2006). Unit stratotypes for global stages: The Neogene perspective. *Earth Science Reviews, 74*, 113–125.

Ilyenkov, E. (1982). *Dialectics of the abstract and the concrete in Marx's Capital.* Progress Publishers.
Kirchner, J. (2002). Evolutionary speed limits inferred from the fossil record. *Nature, 415*, 65–68.
Kirchner, J., & Weil, A. (2000). Delayed biological recovery from extinctions throughout the fossil record. *Nature, 404*, 177–180.
Klein, G. (2015). The "Anthropocene": What is its geological utility? (Answer: It has none!). *Episodes, 38*, 218.
Knox, R. W., Pearson, P. N., Barry, T. L., et al. (2012). Examining the case for the use of the Tertiary as a formal period or informal unit. *Proceedings of the Geologist Association, 123*, 390–393.
Luciano, E. (2022). Is 'Anthropocene' a suitable chronostratigraphic term? *Anthropocene Science, 1*, 29–41.
Malm, A. (2016). *Fossil capital: The rise of steam power and the roots of global warming.* Verso.
Marx, K. (1864). The process of production of capital, Draft Chapter 6 of Capital. MECW 34 (ed) available at: https://www.marxists.org/archive/marx/works/1864/economic/
Moore, J. W. (2017). The Capitalocene, part I: On the nature and origins of our ecological crisis. *The Journal of Peasant Studies, 44*, 594–630.
Moore, J. W. (2018). The Capitalocene, part II: Accumulation by appropriation and the centrality of unpaid work/energy. *The Journal of Peasant Studies, 45*, 237–279.
Oldfield, F., & Steffen, W. (2014). Anthropogenic climate change and the nature of Earth System science. *The Anthropocene Review, 1*, 70–75.
Rockström, J., Steffen, W., Noone, K., et al. (2009). A safe operating space for humanity. *Nature, 461*, 472–475.
Rosental, M. M. (1962). *Principios de lógica dialéctica.* Ediciones Pueblos Unidos.
Ruddiman, W. F., He, F., Vavrus, S. J., et al. (2020). The early anthropogenic hypothesis. *Quaternary Science Reviews, 240*, 106386.
Rull, V. (2017). The 'Anthropocene': Neglects, misconceptions, and possible futures. *EMBO Reports, 18*, 1056–1060.
Rull, V. (2018). What if the 'Anthropocene' is not formalized as a new geological series/epoch? *Quaternary, 1*, 24.
Rull, V. (2020). The Anthropozoic era revisited. *Lethaia, 54*, 289–299.
Saoserov, M. I. (1960) El fenómeno y la esencia. In Rosental & Stacks (Eds.) *Categorías del materialismo dialéctico.* Grijalbo
Schellnhuber, H. J. (1999). 'Earth system' analysis and the second Copernican revolution. *Nature, 402*, C19–C23.
Shanster, E. V. (1973) Anthropogenic system (period). In Prokhorov (Ed.) *Great soviet encyclopedia* vol. 2. Macmillan.

Soriano, C. (2020). On the Anthropocene formalization and the proposal by the Anthropocene Working Group. *Geologica Acta, 18*, 1–10.
Soriano, C. (2021). *Antropoceno, reproducción de capital y comunismo*. Maia.
Soriano, C. (2022). Epistemological limitations of Earth system science to confront the Anthropocene crisis. *The Anthropocene Review, 9*, 111–125.
Soriano, C. (2024). The problems of the Anthropocene in the geologic time scale, and beyond. *Earth Science Reviews, 253*, 104796.
Sprain, C., Renne, P., Vanderkluysen, L., et al. (2019). The eruptive tempo of Deccan volcanism in relation to the Cretaceous–Paleogene boundary. *Science, 363*, 866–870.
Steffen, W., Broadgate, W., Deutsch, L., et al. (2015a). The trajectory of the Anthropocene. *The Anthropocene Review, 2*, 81–98.
Steffen, W., Richardson, K., Rockström, J., et al. (2015b). Planetary boundaries: Guiding human development on a changing planet. *Science, 347*, 1259855.
Steffen, W., Leinfelder, R., Zalasiewicz, J., et al. (2016). Stratigraphic and earth system approaches to defining the Anthropocene. *Earth's Future, 4*, 324–345.
Steffen, W., Richardson, K., Rockström, J., et al. (2020). The emergence and evolution of earth system science. *Nature Reviews Earth & Environment, 1*, 54–63.
Steffen, W., Rockström, J., Richardson, K., et al. (2018). Trajectories of the earth system in the Anthropocene. *PNAS, 115*, 8252–8259.
Trischler, H. (2016). The Anthropocene. *NTM Journal of the History of Science, Technology and Medicine, 24*, 309–335.
Vai, G. B. (2007). A history of chronostratigraphy. *Stratigraphy, 4*, 83–97.
Visconti, G. (2014). Anthropocene: Another academic invention? *Rendiconti Lincei Science Fisiche e Naturali, 25*, 381–392.
Walker, M., Bauer, A. M., Edgeworth, M., et al. (2024). The Anthropocene is best understood as an ongoing, intensifying, diachronous event. *Boreas, 53*, 1–3.
Walker, M., Gibbard, P., & Lowe, J. (2015). Comment on "When did the Anthropocene begin? A mid-twentieth century boundary is stratigraphically optimal" by Jan Zalasiewicz et al. (2015), Quaternary International, 383, 196–203. *Quaternary International, 383*, 204–207.
Waters, C. N., Turner, S., Zalasiewicz, J., et al. (2023). Candidate sites and other reference sections for the Global boundary Stratigraphic Section and Point of the Anthropocene Series. *The Anthropocene Review, 10*, 3–24.
Waters, C. N., Williams, M., Zalasiewicz, J., et al. (2022). Epochs, events and episodes: Marking the geological impact of humans. *Earth-Sciences Reviews, 234*, 104171.
Zalasiewicz, J., Waters, C. N., & Ellis, E. C. (2021). The Anthropocene: Comparing its meaning in geology (chronostratigraphy) with conceptual approaches arising in other disciplines. *Earth's Future, 9*, e2020EF001896.

Zalasiewicz, J., Smith, A., Brenchley, P., et al. (2004). Simplifying the stratigraphy of time. *Geology, 32*, 1–4.

Zalasiewicz, J., Waters, C. N., Head, M. J., et al. (2019). A formal Anthropocene is compatible with but distinct from its diachronous anthropogenic counterparts: A response to W.F. Ruddiman's three flaws in defining a formal Anthropocene. *Progress in Physical Geography, 43*, 319–333.

Zalasiewicz, J., Waters, C. N., Wolfe, A. P., et al. (2017). Making the case for a formal Anthropocene Epoch: An analysis of ongoing critiques. *Newsletter on Stratigraphy, 50*, 205–226.

CHAPTER 3

Structural Link of the Habitability Crisis to the Capitalist Mode of Social Reproduction

The Anthropocene has been characterized as a phase in Earth's history in which humans become a planetary force, capable of driving transformations on a planetary scale similar to those driven by non-anthropogenic natural forces (Fischer-Kowalski et al., 2014). Such a conceptualization has achieved a broad scientific consensus and is consistent with a broader global recognition that human impact on Earth has reached a worrisome level. The empirical evidence of anthropogenic impact on the planet is robust, and what distinguishes the current global environmental degradation from similar scenarios in previous geological periods is, first, its anthropogenic nature, and, second, the velocity and magnitude with which some of the measured parameters are changing.

The Anthropocene has generated a debate that transcends the fields of stratigraphy and geology and has become a topic of global interest with political implications (Swanson, 2016). Researchers from a wide range of disciplines in the natural and social sciences, including biologists, environmentalists, geographers, historians, physicists, economists, philosophers, politicians, etc., are currently engaged in the Anthropocene debate. The reason for this broad interest is that the Anthropocene is not simply a geological question of how it fits into the Geologic Time Scale. Rather, the concept of the Anthropocene has enormous implications for the social organization of humans on Earth, including their own survival as one of

© The Author(s), under exclusive license to Springer Nature Switzerland AG 2024
C. Soriano Clemente, *Marxism and Earth's Habitability Crisis*, Marx, Engels, and Marxisms,
https://doi.org/10.1007/978-3-031-72537-1_3

the species that once emerged as a result of the natural evolution of life. In short, the Anthropocene expresses humanity's preoccupation with the impact of human actions on Earth and asks whether those actions threaten life on Earth. In this respect, whether or not the Anthropocene succeeds as a unit of the GTS is of relative importance because the Anthropocene is here to stay (Zeller, 2015).

If the environmental degradation associated with the Anthropocene has an anthropogenic character, it is related to the way humans interact with nature through their social organization, which evolves and changes with human history. Analytical categories are therefore needed to distinguish the different forms of human social organization in history and their characteristic interactions with nature. Although the empirical data generated by studies of environmental degradation have created a broad consensus in conceptualizing the Anthropocene as a global ecological crisis, the integration of such data into an organic theoretical framework aimed at better understanding the ultimate causes of the crisis is a matter of debate and often strong disagreement.[1] This is somewhat surprising because the empirical data of environmental degradation also provide the basis for clearly limiting the global environmental crisis to a specific period in human history. Great Acceleration charts, species extinction, atmospheric CO_2, global temperature, and many other estimates of the ecological crisis provide a clear empirical correlation with the capitalist mode of social production. These data show that the environmental crisis is intimately linked to a specific socioeconomic organization of human beings, rather than to human beings in general. This allows us to conceptualize the environmental crisis not as some kind of natural and deterministic evolution of humanity, but as a historically determined process. On the basis of simple empirical correlation, it is clear that the global environmental crisis we are experiencing cannot be attributed to the socioeconomic organization of, say, the Romans or the Aztecs. Rather, it is somehow related to the historical configuration of capitalism and the still ongoing expansion of capitalism from Western Europe to the rest of the world that constitutes what we can call modern global society. Human-caused ecological disasters are not exclusive to capitalism and are historically documented in other social formations, such as the extinction of species in ancient Egypt and deforestation in ancient Greece

---

[1] Among the issues debated is the role of the ruling class and fetishism in the habitability crisis (Cunha 2015; Hornborg and Malm 2016).

(Hornborg, 2013). However, this should not be understood as a kind of determinism in the relationship between humans and nature, meaning that environmental degradation is inherently linked to humans. All species have an impact on the Earth, although obviously not to the same extent, but only humans are aware of their impact and have the potential, based on scientific knowledge, to regulate this impact according to the universal metabolism of nature. Environmental degradation under capitalism differs from that under previous modes of production by its planetary scale, by the velocity and magnitude with which it occurs, and by its multifactorial character, that is, the inertia and feedback mechanisms of the processes involved in the crisis of habitability. The empirical correlation does not in itself provide evidence of the ontological link between capitalism and the current ecological crisis. But it does allow us to focus the investigation of the causes of environmental degradation on the nature of the capitalist mode of production, rather than on other modes of production, and thus to understand the deep roots of the ecological crisis.

3.1 Marx's Perspective is Relevant for Understanding the Habitability Crisis

The reasons for adopting a Marxian perspective to understand the Anthropocene and the planetary crisis of habitability are manifold. First, ongoing environmental degradation is a problem that affects the material reproduction of capitalist society, and the more degradation progresses, the more difficulties social reproduction based on the reproduction of capital will have. Therefore, a materialist-based approach, such as that of Marx, which considers not only the natural side of the problem but also the social side—that is, the particular relationship of the ecological crisis to the reproduction of capital—is necessary. Secondly, the social and economic laws unfolded by Marx in Capital are internally articulated within the framework of his labor theory of value, which up to now constitutes the only scientific and organic theoretical corpus we have to understand the capitalist mode of production. Therefore, any research aimed at establishing the structural links between capitalism and the crisis of habitability must be based on Marx's Capital, where his labor theory of value is best exposed. Third, a major challenge posed by the Anthropocene is the need to overcome traditional dichotomies, such as human and nature, and social and natural sciences, which until recently have been maintained as categories corresponding to separate and unrelated fields of knowledge. This is a recurring claim of some scholars of the Anthropocene and Earth system science (Ellis et al., 2016; Guillaume, 2014; Oldfield, 2018; Palsson et al., 2013). Indeed, the Anthropocene has been read as a rupture in Earth system science and, more broadly, as a paradigm shift that transcends the natural sciences (Hamilton, 2016). Overcoming these dichotomies certainly does not mean falling into postmodern relativism, but rather keeping such categories as necessary analytical tools to address their particular relationship at the core of an organic theory.

Marx's approach aims at critically understanding the capitalist mode of production from the point of view of its totality, but emphasizing the essentials of this mode of production that distinguish it from other forms of social organization. In order to understand the fundamentals of capital reproduction, social reproduction must be abstracted from nature, which does not mean that nature is considered a passive actor. On the contrary, as a true materialist, Marx is clear that nature is the ultimate determinant of any form of social reproduction. Abstraction is only a necessary methodological step that allows the subsequent integration of the natural and social realms into a comprehensive and concrete understanding of their mutual interrelation. Such an understanding must be not only materialistic but also dialectical, that is, capable of explaining the movement or evolution of the interaction between humans and nature throughout history and of characterizing the concrete mediations that humans have interposed in their social metabolism with nature. In this regard, the concepts of universal metabolism, social metabolism, and metabolic rupture, which Marx fully integrates into his labor theory of value, become necessary for a scientific understanding of the planetary crisis.[2]

Anthropocene and Earth system theorists have made valuable efforts to characterize and quantify the crisis of habitability in the Anthropocene. However, they lack the appropriate analytical tools and theoretical frameworks to understand the deep roots of the crisis and its connection to social reproduction in general and the production and reproduction of capital in particular. For this reason, scholars of the Anthropocene tend to attribute the crisis to specific technical aspects of social production, such as the production and consumption of fossil fuels, the production of plastics, the use of fertilizers, and to overpopulation and overconsumption

---

[2] These concepts have been unfolded within Marxian theory by researchers of the Monthly Review school. See, for example, Foster (2013).

in general, rather than to the foundations of the particular form of social reproduction based on capital. Such an approach can be characterized as neo-Malthusian, and in this respect, Marx's critique of Malthus and Malthusianism becomes a pertinent perspective.[3] Moreover, no matter how significant fossil fuels, plastics, fertilizers, etc. are for environmental degradation, the mere substitution of these compounds by other less polluting ones becomes useless in the long run unless a change in the social mode of production is undertaken. These mechanistic approaches to the problem of habitability crisis tend to end up in a kind of scientific and technological fetishism, and here Marx's work on alienation and fetishism is very helpful. For example, the reduction of the ozone hole over the Earth's South Pole is claimed to be a success of the 1987 Montreal Protocol, which replaced chlorofluorocarbons with less harmful compounds such as hydrofluorocarbons. However, hydrofluorocarbons are much more harmful to global warming than common greenhouse gases such as carbon dioxide. In fact, the overall planetary crisis has worsened significantly since 1987, despite the reduction of the ozone hole. For all these reasons, a Marxist-based approach is not only appropriate, but necessary if a scientific understanding of the planetary crisis of the Anthropocene is to be pursued.

3.2 Reproduction and Accumulation of Capital Throughout Commodity Production

Marx's epistemological approach based on materialism and dialectics has the potential to provide a better understanding of the Anthropocene and the current ecological crisis and its relation to capitalism. This has been acknowledged by scholars in the Marxist tradition and also by non-Marxist scholars. The literature on Marx is vast; after all, he is considered the most influential scholar of all time (Van Noorden, 2013). In particular, his unfinished masterpiece, Capital, is the subject of extensive scholarly literature and debate and has been read as the dialectical unfolding of his labor theory of value and alienation-fetishism theory. It is beyond the scope of this section to delve deeply into these fundamental

---

[3] For some examples of the neoMalthusian perspectives on the planetary crisis of habitability and on the world population see Kremer (1993), Kapitza (1996), and Ehrlich and Ehrlich (2012).

foundations of Marx's thought. However, a brief exposition of the fundamentals of Marx's theory of value is necessary to show how material and ideal reproduction is undertaken in the capitalist society.

3.2.1 The Material Reproduction of Bourgeois Society

Marx's Capital has been called an ontology of modern society, where modern society is understood as the organization of the world system under the capitalist mode of production (Martínez Marzoa, 1981). Marx's ontology explains the material (economic) reproduction of modern society through of the simple equation $M - C - M'$, where M is money-capital, C is commodity, and $M' = M + \Delta M$. This equation accounts for the production, circulation, consumption, and reproduction of capital, and without capital reproduction, modern society cannot reproduce itself. In capitalist production, the products of labor, labor power, and even labor itself are alienated from the real producers. In other words, workers have to sell the products of their labor and their labor power in the market in order to obtain the necessary means of subsistence. The alienation of labor within the equation of production and reproduction of capital means that the practical life of people, their subsistence and sociability, is mediated by capital in its various forms (commodities, money, finance capital, etc.). Marx recognized that in order to transcend capitalism, it is necessary to examine the specific forms of capital that mediate social reproduction and how these forms are internally intertwined within the global reproduction of capital. For Marx, "all science would be superfluous if the outward appearance and the essence of things directly coincided" (Marx, 1967a, p. 592). Hence, Marx's concept of science requires the dialectical connection of the essence of things with their phenomenal expression across the various levels of abstraction that link the abstract and concrete categories needed to understand the object of study.

In Capital, Marx reveals the dual character of commodities, which consist of use value and value, the latter only manifested as the exchange value of commodities, the phenomenal form of value whose monetary form is price. A commodity is an objectified labor, and in this respect, labor is the only substance of value. The dual character of commodities also expresses the contradiction between concrete labor and abstract labor: as a use value, every commodity is concrete labor capable of satisfying specific needs, and as a value, a commodity is abstract labor, that is, labor reduced to standard homogeneous labor suitable for exchange. Abstract labor is a social property sanctioned only post-festum by the exchange of commodities. While the concrete labor expressed in the use value of commodities is inherently immeasurable, i.e., only the number of commodities as use values can be measured, but not the use value as such, abstract labor can be measured in terms of the time required to produce a commodity. This time, however, is not the actual time used to produce a commodity in a given production process, but the average time that society uses to produce it. This social average time constitutes the magnitude of value, which changes during the historical development of capitalism as labor productivity changes. Therefore, the value of a commodity C is defined by $C = c + v + s$, where c is the value of constant capital (raw and accessory materials, infrastructure, machines), v is the value of variable capital (the portion of capital invested in the labor force) and s is the surplus value obtained from the exploitation of the labor force, that is, the unpaid labor that makes ΔM possible in the capital equation $M - C - M'$.

The different components of capital play different roles in the production process. Labor has the ability to create new value (s) while transferring the value of the constant capital into the value of the commodity. Most raw and auxiliary materials are fully consumed in the production process and their value is fully transferred into the commodity by the living activity of labor. On the contrary, machines and infrastructures are partially consumed, and only the proportional part of the value consumed in the production process is transferred by labor into the commodity. A social production oriented to the production of value, such as capitalist production, means that use values are produced as long as they are the unavoidable carriers of value. In other words, no commodity—the social form of wealth in capitalism—is produced unless it satisfies the condition of the valorization of capital, as shown by $C = c + v + s$ and $M - C - M'$. Thus, Marx's equations of capital's valorization explain the reproduction of bourgeois society through commodity production. Not only must the condition of capital valorization be satisfied, but capitalist production is aimed at obtaining as much s or ΔM as possible. Among other mechanisms, this is essentially achieved by reducing the value of the labor force, i.e. the labor paid by capital, and by accelerating the rotation of capital in $M - C - M'$. Essentially, the condition for the valorization of capital is the law of value, which determines that capital must accumulate in society. The accumulation of capital, however, implies the accumulation of individuals whose labor makes possible the transformation of constant capital (machines, raw materials, etc.) into commodities and the creation of surplus value in the production processes. For this reason, not only capital but also the workers, whose labor is fundamental to the accumulation of capital, must accumulate in society. This leads to an overpopulated world under capitalism, which is a peculiar feature of the capitalist mode of production compared to previous modes in human history.

The law of value applies to the production of all goods and services, not just physical commodities so that the reproduction of all modern society is determined by these capitalist conditions. Value-oriented production is a peculiarity of the capitalist mode of production that distinguishes it from other modes of production. Nevertheless, only use values and not values satisfy human needs, and the fact that under capitalism value is the non-negotiable mediation for any use value to exist shows one of the main contradictions of this mode of production. Namely, human needs are satisfied only when the law of value is satisfied. A social mode of production determined by value and the forms of value—profit, capital, salary, rent, etc.—in which individual labor and the products of labor must be validated *post festum* through the exchange of commodities, and in which the social character of society appears in an alienated form as the sociability of things, must necessarily conform to laws that appear to be determined by things, laws that are established "behind the back of individuals" (Mészáros, 2010, p. 340). These are laws that escape the control of the real producers, such as the law of exchange of labor-value equivalents, the law of capital accumulation, and the law of the tendency of the rate of profit to fall.

3.2.2 *Capital as an Automatic Fetish*

In Capital, Marx explains not only the material reproduction of bourgeois society but also its ideal reproduction, with fetishism based on the objectivity of economic categories (commodities, money, capital, profit, etc.) as the characteristic ideological form of the capitalist mode of production (Gandler, 2006). Fetishism, however, is not an illusion in people's minds but is derived from the real fact that under capitalism social relations between people are mediated by things, i.e. they appear as relations between things, and on the basis of this real perception, things are attributed with properties that correspond to people. Needless to say, the inversion of attributes between humans and things, through which humans are reified and things have an agenda and acquire the status of living things, is carried out by humans themselves. As Marx noted, "we are not aware of this, nevertheless we do it" or, put in other words, we do it in an unconsciously conscious way (Marx, 1967b, p. 49). The concept of fetishism is closely linked to that of alienation, which reaches its maximum expression under capitalism, since social reproduction appears in an alienated form, as a relationship between things that escapes human domination. In Capital, Marx develops the economic categories of bourgeois society from the simplest, concrete and essential form of value, the commodity, to the more complex forms of value that appear on the surface of economic phenomena (wages, rent, credit, fictitious capital, etc.). Marx's development of economic categories goes hand in hand with the development of the ideal forms that constitute the dominant consciousness in bourgeois society, in such a way that the ideal forms are linked to their materialist basis as manifestations of economic categories. The process by which capital becomes a subject, an automatic fetish that subsumes virtually every part of human life, is deeply rooted in commodity fetishism. Marx draws an analogy between the world of commodities and the religious world:

> "In order, therefore, to find an analogy, we must have recourse to the mist-enveloped regions of the religious world. In that world the productions of the human brain appear as independent beings endowed with life, and entering into relation both with one another and the human race. So it is in the world of commodities with the products of men's hands. This I call the Fetishism which attaches itself to the products of labor, so soon as they are produced as commodities, and which is therefore inseparable from the production of commodities" (Marx, 1967b, p. 48).

The commodity is the embodiment of the social reproduction system based on value. As long as capital is a form of value made possible by the concrete mediation of the labor power—a commodity with the unique virtue in the realm of commodities of producing more value than it costs—fetishism is also attached to capital, to which almost magical qualities are attributed. In addition, insofar capitalist production is commodity production, capital is seemingly endowed with the ability to reproduce itself and to accumulate in society apparently without limits. Thus, the expanded reproduction of capital becomes the ultimate goal of capitalist society. The complete fetish form of capital is the financial capital and its movement, which is expressed by Marx as $M\text{-}M'$. Here, the reproduction of capital appears as unplugged from the requirement of commodity production in $M\text{-}C\text{-}M'$, and capital becomes a completely self-referred subject: it is money that generates more money without any mediation. In its fully fetishized form, capital appears as a self-valorized value that is both an end in itself and a self-mediation capable of achieving that end. A contemporary expression of this fetishism is the illusion that central banks can manage the capitalist economy simply by raising and lowering the price of money. In fact, this only works if it is accompanied by coercive measures in society, including repression by police and military corps, which are exercised both in the intra-capitalist economies and from the Western industrialized countries to the former colonies. For these reasons, Marx understood capital as an automatic fetish, a subject with an autonomous agenda that not only escapes the control of human beings themselves but also determines human life in modern society.

Marx's categories can be seen as determinations that are dialectically related to each other in a causal way and that develop in the course of the development of the value-based system of social production towards its fully capitalist form. In other words, Marx's categories are determinations that are determined and are determinants in the dynamic system of social production. In this context, capital can be seen as the supradetermination that determines the social production and reproduction in bourgeois society. Throughout the historical process of capital's expansion and consolidation worldwide, capital presents itself as the precondition, mediation, and result of social production and reproduction. However, capital is only self-mediating to the extent that labor is treated as a part of itself, as a mere means for its advancement, and it is here that all the contradictions of capitalist production lie. In reality, only labor is the universal practical activity that allows humans to be humans and allows the evolution of humanity, while capital is only an alienated historical mediation and not a human universal (Engels, 1986).

In the capitalist production, the relationship between labor and capital appears as an irreconcilable antinomy. On the one hand, capital, a product of labor, utilizes the labor power in the production process as an alien mediation whereby capital is able to obtain surplus value, and thus reproduce itself. On the other hand, for the real producers dispossessed of the means of production, their labor power is useless without the mediation of capital, which is seen as a mediation alien to labor. In this way, techno-scientific progress, which is based on the accumulated social labor culturally inherited through human generations, is reduced to a means of obtaining as much surplus labor as possible. As a result, the universality of labor and its ability to produce use values that satisfy human needs is subsumed under the automatic subject of capital to serve its own self-mediation and self-valorization. Thus, a social relationship that is effectively carried out by people, and that is mediation between people, becomes mediation *of* and *for* capital, and in this way, capital becomes an autonomous subject empowered over humans. For these reasons, in the process of capitalist production individuals are personifications and reifications of economic categories: capitalists are personifications of capital, while workers are reduced to mere things or costs of the production process. The accumulation and reproduction of capital engender the tendency for the rate of profit to fall and the alienated form of social metabolism with nature, which constitute examples of how the automatic fetish of capital conducts the process of social reproduction, and of the exacerbation of the contradictions of the system that have ultimately led to the ongoing crisis of habitability on Earth.

3.2.3 *The Tendency of the Rate of Profit to Fall*

Based on empirical observation and logical deduction from the laws of capitalist production, such as the law of value and the law of capitalist accumulation, Marx concluded that the rate of profit has a tendency to fall in the long-term history of capitalist production. According to Marx, the tendency of the rate of profit to fall is one of the most important laws of political economy because it expresses the fundamental contradiction of the capitalist mode of production, that is, the contradiction between labor productivity and the social relationship between labor and capital. In other words, the contradiction between the development of the productive forces and the relations of production that make this development possible in capitalism. Marx not only explained the essence of the law but also detailed a number of countertendencies activated by capital to mitigate the fall of profit (Marx, 1967a, p. 149–178).

The decline in the rate of profit had already been observed empirically by nineteenth-century political economists, who did not succeed in explaining this phenomenon. Today, some bourgeois economists, such as Nobel laureate Paul Krugman, implicitly recognize the problems of robotics, i.e., the substitution of living labor by dead labor, for the valorization of capital (Krugman, 2012). However, like the political economists of the nineteenth century, today's bourgeois economists are not equipped with the appropriate theoretical tools to understand the rate of profit and its tendency to decline within the framework of an organic theory of the production and reproduction of capital.

Capitalists are personifications of capital, and from their perspective, the value of a commodity $C = c + v + s$ is expressed in terms of the cost-profit equation $C = k + p$, where $k = c + v$ is price cost and $p = s$ is profit, the phenomenal form of surplus value. This equation expresses the reality of capitalist production as seen by capitalists and states that the more money is invested in production, the more profit is made. The ratio between profit and the total capital invested in the production process is the rate of profit $p' = s / (c + v)$. Dividing the former expression by v we get $p' = (s/v) / [(c/v) + (v/v)]$, which expresses the rate of profit in terms of the rate of surplus value, $s' = s/v$, and of the organic composition of capital, $q = c/v$, which is the ratio between the value of constant capital and the value of the labor force. Simplifying, we get $p' = s' / (q + 1)$. Individual capitals try to extract as much p as possible from the production process with the lowest possible capital investment. This is usually achieved by introducing technical-scientific progress, which provides a competitive advantage over other capitalists because it allows them to produce more goods than their competitors, which can be sold at lower prices. Capital may also migrate to economic sectors with higher profitability in order to increase profit. With the introduction of techno-scientific advances labor becomes more productive but the ratio between constant capital and labor force, q, increases or, in other words, less labor is needed to produce more commodities. Sooner or later, however, the techno-scientific advances first introduced in some economic sectors extend to other sectors of social production, and this leads to an increase in the organic composition of society's global capital. As a result, more means of production or constant capital, c, and less labor force or variable capital, v, are progressively incorporated in the production processes of society. However, only the labor force and not constant capital is able to produce value or, put in Engels' words:

> "The law of value is aimed from the first against the idea derived from the capitalist mode of thought that accumulated labour of the past, which comprises capital, is not merely a certain sum of finished value, but that, because a factor in production and the formation of profit, it also produces

value and is hence a source of more value than it has itself; it establishes that living labour alone possesses this faculty" (Marx, 1967a, Engels' Preface).

The substitution of living labor for past or dead labor increases in the historical course of capitalist production and has the long-term effect of lowering the rate of profit. Moreover, the more living labor is substituted, the less living labor remains to be substituted and thus produce surplus value. This is the main contradiction of capitalist production, as shown by the economic phenomenon of profit: labor becomes more productive, i.e., the quantity of commodities produced by a unit of labor v is greater, and certainly the global sum of profit or surplus value is also greater, but on the other hand, each individual commodity contains less surplus value, and the rate of profit decreases. As a result of the law of the tendency of the rate of profit to decrease, an increase in labor productivity today is a decrease in the rate of profit tomorrow, and that is why the law is best expressed in the long term of capitalist production. Although the law of the decreasing rate of profit has been contested by Marxist and non-Marxist scholars, today, the empirical evidence for a decreasing rate of profit in the long-term history of capitalist production is very robust, in particular for capitalist developed countries and after the Second World War (Basu et al., 2022; Carchedi & Roberts, 2018; Freeman, 2019; Smith et al., 2021; Zachariah, 2009). The main reason for this robustness in advanced capitalist economies is that bourgeois States and international organizations have notably improved the national and global accounting of the economy. Nevertheless, there is an increasing number of studies with estimates of the rate of profit for historical periods spanning nearly a century, not just the second half of the twentieth century (Jones, 2016; Maito, 2016).

Faced with the tendency of the rate of profit to fall, capital activates mechanisms that aggravate the causes of this fall. Essentially, by increasing the substitution of living labor for dead labor, which increases the organic composition of capital and thus labor productivity. Capital can also mitigate the fall in the rate of profit for short periods of time through a number of other mechanisms, such as the reduction of wages, the extension of the working day, and the intensification of labor, all of which are aimed at increasing the rate of surplus value obtained from production. Extending the working day and intensifying labor were also characteristic forms of exploitation of slave labor in ancient regimes, and trimming wages could be somewhat similar to increasing taxes on bonded labor in feudal regimes. In any case, these mechanisms of labor exploitation are limited by absolute physical or human constraints.

The specific form of labor exploitation under capitalism that ensures an increase in the rate of surplus value is the devaluation of the labor force. This is a novel mechanism of labor exploitation that distinguishes the capitalist mode from any other historical mode of social production based on labor exploitation. However, as living labor is increasingly expelled from the production processes, an increase in the rate of surplus value cannot, in the long run, compensate for the increase in the organic composition that capital must undertake to increase labor productivity. The long-term effect of increasing productivity on a capitalist basis is a further decrease in the rate of profit, more and new commodities have to be produced to overcome such a decrease, and more capital has to be invested, so that capital overaccumulates in society. Thus, the way capital faces the falling rate of profit shows it as an automatic fetish, engaged in a blind dynamic aimed at overcoming the internal limit of capital production and reproduction expressed by the tendency of the rate of profit to fall. The irrationality of such a blind dynamic is shown by the fact that, although the more labor productivity is developed, the lower the rate of profit, capital develops labor productivity as its immanent mechanism to overcome the falling rate of profit, thus approaching the internal limit of capitalist production. This expresses the essence of capital as a subject beyond any control, under the command of which people do not know what they are doing, but do it anyway. On a simple intuitive basis, it becomes clear that social reproduction under capital accumulating at a decreasing rate of profit is unsustainable and that the contradictions of capitalist production are insurmountable within this mode of production.

3.2.4 Social Metabolism with Nature and Metabolic Rift

Marx understands humans as part of nature, as one of the species that emerged during the evolution of natural history. For Marx, humans are self-producing social beings and labor is the practical activity that makes this self-production possible:

> "Men can be distinguished from animals by consciousness, by religion or anything else you like. They themselves begin to distinguish themselves from animals as soon as they begin to produce their means of subsistence, a

step which is conditioned by their physical organisation. By producing their means of subsistence men are indirectly producing their actual material life. The way in which men produce their means of subsistence depends first of all on the nature of the actual means of subsistence they find in existence and have to reproduce. This mode of production must not be considered simply as being the production of the physical existence of the individuals. Rather it is a definite form of activity of these individuals, a definite form of expressing their life, a definite mode of life on their part. As individuals express their life, so they are. What they are, therefore, coincides with their production, both with what they produce and with how they produce. The nature of individuals thus depends on the material conditions determining their production" (Marx & Engels, 1976, p. 6).

Although humans are self-mediated beings of nature, they cannot completely detach themselves from nature and are always bound to the laws prescribed by nature. Like other species, human beings engage in a matter exchange or metabolism with nature. But for humans, this metabolism is strongly social and mediated by labor, which gives human social metabolism a teleological and conscious character. Thus, social metabolism is not only a physical exchange, but also implies a complex interaction with nature through which humans have built a human world that is both material and ideal, and that is socially inherited throughout human history. Social metabolism with nature is common to all socioeconomic organizations in human history, and labor as the practical mediation of this social metabolism is labor in general, the universal, material, and ideal activity by which humans regulate their metabolic interaction with nature. However, the social metabolism between humans and nature takes different forms under the different social organizations of humans and under the various historical forms of labor and labor exploitation in the social organizations.

As for any other species, human social metabolism also includes the reproduction of individuals, that is, the number of individuals that nature can sustain on the basis of available resources. However, humans are capable of transforming nature through labor, and in this way, they are able to transcend the quality and quantity of resources strictly available in nature. Since the human social metabolism with nature is historically determined, so is the human population and population dynamics. For this reason, in Marx's critique of Malthus, instead of a universal population law common to humanity and to all modes of production, the human population is historically determined and each mode of production has its own population law. For example, the social metabolism in ancient modes, in which slave labor is exploited through relationships of property among people, has specific characteristics that differ from those of feudalism, in which bonded labor is exploited through the ownership of land, and both differ from capitalism, based on wage labor and the ownership of the means of production. All these socioeconomic forms have, accordingly, particular population laws and dynamics.

In the capitalist mode of production, the productivity of labor reaches its maximum development in history, while the two sources of all wealth, nature and labor, are objectified as mere things of a production process aimed only at the self-valorization of value (capital). According to Marx, in modern industry and large-scale agriculture "Capitalist production, therefore, develops technology, and the combining together of various processes into a social whole, only by sapping the original sources of all wealth—the soil and the labourer" (Marx, 1967b, p. 330). The reduction of labor and nature to mere objects of the process of social production and reproduction of capital is the basis of the alienated form of social metabolism that characterizes capitalism. This alienated form of social metabolism means that human beings do not consciously control their exchange of matter with nature, but rather that social metabolism is controlled by capital as a blind power and by the economic laws of capital reproduction that take place behind the backs of individuals. Thus, in the alienated form of social metabolism commanded by capital, human beings are deprived of the teleological and conscious character that labor gives to their living activity, including their social metabolism with nature. In capitalist production, capital is the alienated and historical second-order mediation superimposed on labor, the first-order and universal mediation of human social metabolism with nature. In fact, capital is not a mediation for human beings to carry out their social metabolism with nature, but human beings and nature are mediations for capital to carry out its reproduction and its own metabolism.

The alienated form of social metabolism in capitalism necessarily leads to a rupture or rift with the natural laws prescribed by the universal metabolism of nature. Large-scale industry and agriculture, overpopulation concentrated in megacities, and the rupture between life in megacities and life in the countryside, where the land is conceived as a mere dumping ground and a source of food, energy, and raw materials exemplify better than anything else the metabolic rift immanent to the capitalist production. The metabolic rift as exposed in Marx's Capital was mainly conceived as a rupture in the nutrient cycle between the soil and the cities. Based on the research undertaken by organic chemist Justus von Liebig, Marx concluded that the soil is deprived of nutrients, which are concentrated in the cities for food and clothing, while the cities do not return these nutrients to the soil, but instead return pollutants. In this way, the conditions for sustained soil fertility are violated. Since Marx's time, the metabolic rift has extended and intensified to involve many other parts of nature, such as the deep substrate, the atmosphere, the biosphere, the hydrosphere, in short, most of the Earth's geospheres. In this respect, the metabolic rift has globalized and now affects most aspects of the social metabolism between humans and nature. In terms of Earth's habitability, the universal metabolism of nature has been threatened several times along Earth's history, during the known mass extinction episodes. The present mass extinction, however, runs two orders of magnitude faster than the fossil mass extinctions and, thus, its potential magnitude is expected to be greater than that of the fossil ones. As is often said, this changes everything.

Marx understands humanity as the collective social being that evolves throughout history along with the evolution of the practical transformation of nature through labor, and in this regard, nature is the inorganic or extended body of humanity (Marx, 1959). Humans incorporate more and more elements of nature as their inorganic body, and this is a significant quantitative departure from the kind of interaction with nature undertaken by animals or plants, which is limited in scale. The interaction and metabolism of animals and plants with nature is also social, like that of humans, but it is qualitatively different because it is not based on labor. Animals in particular do have a certain teleology in their practical interaction with nature. However, this is an immediate kind of teleology, a kind of direct action-reaction behavior that is not mediated by the abstraction and conceptual elaboration that allow humans to construct a material and ideal world that is culturally inherited throughout history. Therefore, whatever we call the practical interaction of animals with nature, it is qualitatively different from the practical interaction of humans based on labor. Today, humans interact and modify not only the whole Earth but also other planets which have also become the inorganic body of humanity. The development of the metabolic rift, as conceptualized by Marx in the nineteenth century, into today's Anthropocene crisis, or the crisis of the Earth's habitability, is the expression of the contradiction of humanity with nature, with its inorganic body, a contradiction that is historically determined and insurmountable within the limits of capitalist production.

3.3 Implications of the Rate of Profit for the Capitalist Systemic Crisis and the Habitability Crisis

In order to investigate the structural relationship between the capitalist mode based on the reproduction of capital and the planetary crisis in terms of causality and necessity, we must turn to the scientific approach par excellence to this mode of production, that is, to Marx's labor theory of value. To date, none of the proposed alternatives to Marx's theory of value can be considered truly scientific. A scientific theory is a logical-conceptual framework composed of laws or principles that allows understanding of the phenomena observed in the studied system in a way that phenomena are linked throughout concrete mediations to the general and essential laws of the system obtained by abstraction. The concepts and categories that make up the theory are correlated with the objects of the studied reality and are internally intertwined within an organic and dynamic corpus in which they fulfill a specific function that allows the predictive interaction with the studied reality and where practice stands as the ultimate criterion of validation (Ilyenkov, 1982). Examples of scientific theories are plate tectonics, where, for example, oceanic crust subducts under the continental crust and not vice versa; cell theory, where mitochondria perform the respiratory function and no other; particle physics, where electrons necessarily orbit the atomic nucleus; and Marx's theory of value, where commodity, surplus value, capital, and other forms of value are internally articulated under specific roles.

All these theories are scientific because they explain the origin and evolution of the systems under study, because they allow us to understand the history of the systems through the development of the concrete relations that mediate between general laws and singular observable phenomena, and because they are confirmed as approximately correct by practical interaction with the system.[4] This does not mean, however,

---

[4] It may seem aberrant, but the development and subsequent use of the atomic bomb in the Second World War is a practical evidence of the fact that our knowledge of the atomic

that scientific theories are conclusive, and in fact, this is not the case for the above theories and for any scientific theory in general. On the basis of the contradiction between the concepts of the infinite and the finite, Engels unfolded the impossibility of any conclusive knowledge of nature and society:

> "But an adequate, exhaustive scientific exposition of this interconnection, the formation of an exact mental image of the world system in which we live, is impossible for us, and will always remain impossible. If at any time in the development of mankind such a final, conclusive system of the interconnections within the world – physical as well as mental and historical – were brought about, this would mean that human knowledge had reached its limit, and, from the moment when society had been brought into accord with that system, further historical development would be cut short – which would be an absurd idea, sheer nonsense" (Engels, 1947, p. 21).

Marx highlighted the tendency of the rate of profit to fall as one of the most important laws of his theory of value. This capitalist law, together with other laws such as the law of capitalist accumulation, the law of capitalist population, and the law of value forms the organic architecture of the labor theory of value. Nevertheless, the rate of profit and its decreasing tendency in the long historical course of the capitalist mode of production have been the subject of an intense debate within Marxist economic theory for more than a century, the historiography of which cannot be dealt with here. A debate that concerns the theory of crises, the role of the falling rate of profit in the economic crises that periodically shake this mode of social production, and whether the tendency of the rate of profit to fall indicates the long-term historical transience of the capitalist mode. Together with the debate on the transformation of values into prices, the so-called transformation problem, the debate on the rate of profit and the theory of crises constitute the main challenge to Marx's theory of value. It is a debate that goes beyond Marxism and economics and has deep epistemological and philosophical roots regarding the way social reproduction should be organized and the way we understand reality.

---

structure of matter is approximately correct. This knowledge does not necessarily imply, of course, to manufacture nuclear weapons, just as knowledge of the genetic structure of living matter does not necessarily imply cloning organisms.

In the following analysis, the interrelationship of the falling rate of profit with the periodic economic crises of capitalism and with the secular or systemic crisis of capital reproduction is first unfolded. Second, questions are raised about the interpretation given by some Marxists to the tendency of the rate of profit to fall, to Marx's theory of crises, and about how these issues are framed within Marx's theory of value. In particular, the discussion focuses on the debates that have arisen since the outbreak of the 2007–2008 financial crisis and on the epistemological and methodological reasons underlying the critique of Marx's law and Marx's theory of crises. Third, the planetary crisis is conceived as a structural necessity of the capitalist mode of social production in light of the tendency of the rate of profit to decrease.

3.3.1 Declining Rate of Profit, Economic Crises, and Capitalist Systemic Crisis

The tendency of the rate of profit to fall in the historical course of capitalist production is by no means Marx's discovery. Classical political economists such as Adam Smith, David Ricardo, and John Stuart Mill were fully aware of this tendency, although they did not understand it in its full dimension and in relation to the labor theory of value. Nevertheless, the classical political economists sensed the potential danger of a falling rate of profit not only for the healthy development of capitalist production, but also for the long-term sustainability of this mode of production.[5] Only Marx was able to integrate the tendency of the rate of profit to fall with the other laws of capitalist production within the consistent organic framework of the labor theory of value. For Marx, the law of the falling rate of profit is a constitutive part of his labor theory of value, it is a phenomenal form of the law of value and the law of capitalist accumulation, and its dismantling requires the dismantling of the entire conceptual scaffolding of the labor theory of value (Kliman et al., 2013). Marx's theory of value would not be a scientific theory if it did not allow for a thorough understanding of the concrete phenomena inherent in the reproduction of capital, such as the declining rate of profit and periodic

---

[5] "But the main thing about their horror of the falling rate of profit is the feeling that capitalist production ... is not an absolute mode, moreover, that at a certain stage it rather conflicts with its further development" (Marx 1967a, p. 168). Here, the *horror* is that of the classic political economists in the face of the falling rate of profit.

economic crises. By cutting off this theory from the concepts and laws that allow us to explain these and other economic phenomena, we are in fact cutting off the scientific character of the theory itself. Scientific rigor demands that whoever does this provides a positive alternative to Marx's theory of value. This has not yet happened.

Marx's theory of value is based on the fundamental premise that labor is the only value-creating substance: "The labour, however, that forms the substance of value, is homogeneous human labour, expenditure of one uniform labour power" (Marx, 1967b, p.29). Truly, labor is much more than the substance of value, it is the universal practical activity that mediates the metabolic exchange between humans and nature and that has driven human's evolution (Engels, 1986, see The Part Played by Labour in the Transition from Ape to Man). Because of the teleological character of labor, humans develop the capacity to know and discern ends and means and to transform the natural world around them as they transform themselves. Labor is thus the material activity that makes the materialist conception of ethics and knowledge possible. The creation of the human material world occurs in accordance with the creation of an ideal world that gives meaning to things. It is the world of culture that every human being is confronted with at birth, that is both established and changed throughout history, and that is inherited socially rather than biologically throughout human evolution.[6] Nothing similar exists in the animal or plant kingdoms, and from this perspective any claim that animals, plants, or nature in general create value is meaningless.[7] Strictly speaking, animals and plants work, but they do not labor as humans do. The fact that in some languages, such as Spanish, the same term is used for essential human activity, the activity of animals and plants, and a magnitude of physics, only tells us about the lexical limitations of verbal language as a formal expression of thought, and about the relevance of context in understanding the meaning of terms.

The living labor consumed in the direct process of production sets in motion the dead or past labor objectified in the instruments of labor and the means of production. In this way, living labor transfers all or part of the value of dead labor to the new product and adds surplus value,

---

[6] Soviet philosopher Evald Ilyenkov provides one of the most accurate contributions to the relationship between the material and the ideal, see Ilyenkov (2012).

[7] On the confusion of the so-called ecological economics regarding value and value creation, see Foster and Burkett (2018), and Pigmaier (2021).

the "converted form" of which is profit (Marx, 1967a, p. 27). The rate of profit decreases as living labor is increasingly replaced by dead labor in the production process. Moreover, the more living labor is replaced by dead labor, the less living labor remains to be replaced and the less profit it can generate in relation to the capital invested. The interposing of labor instruments, a world of objects, is certainly an intrinsic and universal feature of the metabolic relationship between humans and nature. In fact, human beings evolve because they interpose this material-ideal mediation in their relationship with the world, a mediation that is inherited and accumulated in the course of human history. With the capitalist mode of social reproduction, there is an enormous development of knowledge about the ends and means of the processes of production. The capitalist mode entails the development of the modern system of science and technology, on the basis of which the substitution of living labor for dead labor undergoes an unprecedented acceleration in the history of mankind. The consequences of this substitution for a system of social reproduction based on the valorization of value, in which living labor, the only element capable of creating new value, is increasingly excluded from the production of value, are dramatic.

Profit is the ultimate goal of the production process for each individual capital, and the various capitals compete for a share of the total profit produced in society through the exploitation of the labor of the working class. The competition for profit at a decreasing rate forces individual capitalists to increase their investment of capital and the number of commodities produced in order to obtain an equal or greater share of the total profit produced in society than that of other capitalist competitors. This dynamic leads to the overaccumulation of social global capital and to the overproduction of commodities. Overaccumulation and overproduction acquire a chronic character with the historical development of capitalist production. The overproduction of commodities is also expressed as underconsumption, as goods produced in excess of social consumption needs, which include consumption in productive processes and individual consumption of people, which are increasingly difficult to sell. The capitals, competing for profit with a declining rate, initiate a series of mechanisms aimed at mitigating the fall of this rate. For example, a faster rotation of the reproductive cycles of capital allows an increase in the surplus value; the increase in working hours and the intensification of work rhythms have the effect of increasing the surplus value obtained by capital, and thus the profit, in relation to the capital invested; the increase in the exploitation of labor, or the rate of surplus value, also fulfills this role. This later mechanism is twofold. In terms of labor, the necessary labor time with which workers pay for their means of subsistence is transformed into surplus labor time. In terms of value, the increase in the rate of surplus value reduces the value of the labor that capital must pay in relation to the surplus value it receives. However, the mechanisms aimed at mitigating the fall in the rate of profit suffer from intrinsic limitations that hinder their long-term effectiveness. Increasing the rotation of the reproductive cycles of capital usually entails higher capital investment, especially circulating capital, and therefore its effectiveness in mitigating the fall in the rate of profit is limited. The transformation of necessary labor time into surplus labor time cannot be carried to the extreme, where all labor time is surplus labor time, because the working class would not be able to pay for its means of subsistence, which are also a source of surplus value for capital. The cheapening of labor depends on the increase in productivity, including the productivity of those sectors that produce the means of subsistence, and therefore its effectiveness in counteracting the fall in the rate of profit is also limited. In fact, both the length of the working day and its division into necessary labor time and surplus labor time must be balanced in such a way as to allow: (a) sufficient health conditions for the working class to work day after day; (b) minimum necessary labor time to pay for the means of subsistence of the working class; (c) sufficient time outside of working hours to consume the increasing number of commodities produced, which are sources of capital valorization. Thus, the daily time, including the length of the working day and its division into necessary labor time and surplus labor time, must be balanced in such a way as to allow the reproduction of the working class as a laboring class, and this sets a physical limit to the countertendencies that capital raises to mitigate the fall in the rate of profit.

Working together and separately, the former mechanisms increase labor productivity, that is, the number of commodities containing shares of the total surplus labor or profit produced. They can temporarily mitigate the fall in the rate of profit, as has happened during a significant part of the neoliberal period. Neoliberalism has been characterized by an enormous migration of capital to the financial sphere, which does not produce value but derives its income from the extraction of the value produced in the sphere of productive capital. In other words, the income of financial capital is the interest on the loan of money that will operate as capital in the productive sector. Capital migrates to the financial sphere because its revenues are based on the payment of debts sanctioned by legal agreements that are, in principle, less dangerous than the profits obtained from the productive sphere. Moreover, by migrating from the productive sector to the financial sector, capital does not have to face directly the social pressure of the working class for higher wages. However, this dynamic only postpones the creation of value, which ultimately sustains any profit and income. The financialization of the economy requires innovative credit and financial engineering mechanisms and even the need for central banks to issue money that is not backed by value.[8] As a result, the fall in the rate of profit has been partially mitigated during neoliberalism, but society has become increasingly indebted.

The commodity system based on value reaches its full form in the capitalist mode of social production based on the reproduction of capital. It is a social system consolidated in a historical process in which individuals and groups of individuals interact spontaneously to produce the material needs for social reproduction. In this system, the interpersonal relations between people are objectified as relations between things, between mercantile products, beyond the consciousness of the individuals themselves. Here, the invisible hand of the market and the system of prices spontaneously built around a general rate of profit are necessary mechanisms for the reproduction of capital and for the navigation of individual capitals competing for a share of the total profit produced in society in a system that appears as chaotic (Cockshott & Nieto, 2017). In such a mode of production, as it becomes increasingly difficult to invest overaccumulated capital on a profit basis and to sell overproduced commodities, economic crises emerge as the natural and ungoverned mechanism that partially restores the rate of profit and the healthy conditions for capital to continue reproducing. Every crisis destroys forms of overaccumulated capital: variable capital, when it excludes many people

---

[8] Traditionally, central bank currency issuance was backed by gold reserves, which were ultimately backed by value. Since 1971, when the US Federal Reserve unilaterally abandoned the Bretton Woods Agreements of 1944, the Federal Reserve has increased the issuance of Dollars backed by less and less gold reserves and value. As a result, the U.S. dollar has become the world's fiduciary currency and the United States the world's largest debtor. More recently, the Central Bank of the European Union has embraced the debt dynamic of issuing currency on the basis of future value. The so-called Modern Monetary Theory (MMT) is, in essence, nothing more than an ideological legitimization with a scientific veneer of the practice of issuing money with no redemption value. For one of many critiques of modern monetary theory, see Del Rosal (2019).

from the production processes and forces them to live on the margins of capital reproduction; commodity capital, which remains unsold and useless; constant capital, with the failure of many companies, their infrastructures and machines. The processes of concentration and centralization of capital, which accompany the reproduction of capital and are intensified in the periodic crises of capitalist production, also serve to reconfigure the size and number of capitals that survive each crisis.

Beyond the specific conjunctures of each crisis, they all involve the destruction of capital and occur on the structural undercurrent of a declining rate of profit. A local crop failure, for example, can certainly trigger a global crisis because of the close interconnectedness of production processes on a global scale. However, this is not the characteristic form of capitalist crises, because it also occurred in earlier modes of production, albeit on a local rather than a global scale. Moreover, if such a bad harvest is related to, say, global climate change, it remains to be understood whether global change is related to the capitalist mode of production or not. For these reasons, the crisis of capital overaccumulation is the characteristic form of crisis in capitalism (Gill, 2002). Most capitalist crises erupt in the financial sphere, where capital has migrated and deceptively decoupled itself from the law of value, from the cumbersome physicality of producing commodities under a declining rate of profit, and where capital has become the self-referential fetish whose movement is M-M'.[9]

Crises erupt at the moment of capital circulation when value must be effectively materialized as money and payments for the value of capital must be confronted. In crises, the mechanisms of credit expansion that capital itself has developed to promote its reproduction, the structure of financial engineering that has led to the financial hypertrophy of the economy, and economic bubbles of various kinds collapse, and the crisis appears as a lack of money when, on the contrary, there is an excess of money. On the surface of economic phenomena, fictitious capital and financial capital appear to have become detached from the production of goods, thus negating the law of value. However, crises are the moment of truth when the law of value appears on the surface of the economy and value must be materialized as money. From this perspective, the articulation of crises in the financial sphere and in the productive sphere is seen

---

[9] For an empirical expression based on the profit rate of the apparent decoupling of financial capital from productive capital, see Freeman (2013).

in the opposite way to how events unfold and how they are conceived by bourgeois economists, in an immediate and unscientific way. For them, and in accordance with the temporal sequence of events, the financial crisis triggers the productive crisis because the first precedes the second, while for Marx it is the opposite.

According to Marx, the financial crisis is a form or manifestation of the underlying crisis in the productive sphere, which manifests itself on the surface of the economy after the financial crisis. This is because the financial sphere is the weakest and least essential sphere of economic architecture, in which money, a form of value, basically functions as a sign of value, seemingly divorced from its internal connection to value. In epistemological terms, Marx operates with the logical categories of necessity, essence, phenomena, and causality from a dialectical point of view that allows understanding crises as a necessity for the reproduction of capital due to its own internal contradictions. In contrast, bourgeois economics operates from a non-dialectical logic and understands crises as accidents of the reproduction of capital due to various conjunctures, because structural contradictions are not allowed, or only at the phenomenological level where they can be formally resolved. After crises, the reconfiguration of capital and the partial recovery of the rate of profit can allow a new period of economic growth in which capital reproduces under healthier conditions than before the crises. However, this new period of economic growth is based on the same premises, the same structural determinations and contradictions that caused the previous crisis. That is, the capitals that survived the crisis compete for a share of the global profit with the same means that caused the crisis, by increasing the magnitude of invested capital and the substitution of living labor by dead labor. The rate of profit may partially recover through a number of different mechanisms: the expansion of capital into new markets, the commodification of unexplored aspects of human life, and the overexploitation of labor through the mechanisms of absolute and relative surplus value. But all these mechanisms working together and separately are based on the same premises that led to the fall of the rate of profit and that propel the general drive of capital to overaccumulate under every and all circumstances.

Thus, the mechanisms underlying each crisis are not only still operative in the new economic periods of capital accumulation after the crises, but they operate on a progressively weakened substrate of capital reproduction, a substrate that will be weakened again in the new crisis yet to come. Therefore, in the long-term historical path of capitalism, the achievement of profit through the exploitation of labor faces increasing difficulties, and thus the whole system of capital reproduction weakens. The state of contemporary capitalism is one of financial hypertrophy and low average rates of profit, although the profit shares of some monopolistic companies can be extraordinary, associated with the long-term fall in the rate of profit, which has only been partially mitigated during neoliberalism. This is a secular or systemic crisis of the valorization of value, punctuated by economic cycles of capital accumulation and recurrent crises, which has also been typified as twilight capitalism, terminal capitalism, and senile capitalism. A scenario in which the production of value is an end in itself alien to social needs, but in which the continued valorization comes up against the ultimately insurmountable obstacle of the expulsion of labor, the only element that creates value, from the process of valorization of capital. The tendency of the rate of profit to fall, together with Marx's theory of crises, points to the main contradiction of the capitalist mode of production based on the reproduction of capital. This is the contradiction of capital with its *raison d'être*, labor, which is at the same time the antagonistic being of capital, and in the end, is the contradiction of capital with itself. In accordance with the blind character of capitalist production, capital escapes forward to face its fundamental contradiction, thus deepening the systemic crisis along the downward slope of the rate of profit. For these reasons, Marx understood capitalism as a mode of production that rules people instead of being ruled by them.

3.3.2 *Revision of the Marxist Debate on the Rate of Profit and the Theory of Crises After the 2007–2008 Financial Crash*

The debate on the rate of profit and its downward trend gained new momentum in the years following the outbreak of the Great Recession of 2007–2008. The main issues of this debate have been: the validity of the law of the falling rate of profit, its relation to capitalist economic crises, whether Marx really had a theory of crises, and whether Marx abandoned this law late in life. David Harvey and Michael Heinrich are the main Marxist scholars who argue against the contemporary relevance of the law, narrowly conceived, and its relation to contemporary capitalist economic crises.[10] Harvey and Heinrich are well-known Marxists whose contributions have been documented elsewhere and need not be emphasized here. The focus will therefore be on their specific critiques of Marx's conception of the rate of profit and crises. In sum, these critiques are based on epistemic paradigms that are arguably antagonistic to those of Marx, and they emphasize a presumed excessive interference by Engels in the compilation and editing of Marx's manuscripts of Volumes II and III of Capital.[11]

3.3.2.1 Michael Heinrich's Critique of the Profit Rate and Crises Theory in the Context of the So-Called New Reading of Marx

The Marx-Engels Gesamtausgabe (MEGA) is a project to publish a historical and critical edition of all the writings of Marx and Engels at all levels of their development (Fineschi, 2008). This is an absolutely necessary and laudable project that began in the early years of the Soviet Union and will allow for the historical and philological reconstruction of all of Marx and Engels' writings. Similar projects to MEGA have been undertaken with natural scientists such as Charles Darwin and Charles Lyell. The new reading of Marx arises in the context of the new edition of MEGA. The historical and philological reconstruction of the writings of a scientific theory can shed light on the process of conceptualization and elaboration of the theory. However, the validity of a theory must be assessed on the basis of the theory itself, taking into account the internal consistency of the theoretical corpus, including the articulation of the elements of the theory, and its ability to explain the phenomena under study. In general, the critique of a scientific theory cannot be based solely on the historical unfolding of the verbal exposition of the process of elaboration of the

---

[10] Comments and replies of the debate between Heinrich and other Marxist scholars can be followed from: https://mronline.org/2013/12/01/heinrich-answers-critics/, see also Kliman et al. (2013). Comments and replies of the debate between Harvey and other Marxist scholars can be followed from: https://marxisthumanistinitiative.org/economics/harvey-vs-marx-on-crisis-klimans-rejoinder.html and https://thenextrecession.wordpress.com/2014/12/17/david-harvey-monomaniacs-and-the-rate-of-profit/, see also Callinicos and Choonara (2016).

[11] In western Marxism, there is a long tradition of blaming Engels for some undesired practices of Marxism, which has been characterized as a Marxist theodicy (Piedra Arencibia, 2019).

theory. Inconsistencies in verbal expression are only of relative importance, and they usually reflect the sometimes accidental and laborious work in the ongoing process of the theory's conceptualization. Criticism based on such inconsistencies should therefore be read with caution.

The new reading of Marx is claimed to be rigorous because it adheres to the literalness and temporal sequence of texts and quotations that serve as the basis for interpreting Marx's theoretical thought. In this way, however, the interpretation of the texts by emphasizing the internal coherence of the content as a whole and beyond its particular verbal exposition in the various texts and fragments can be somewhat neglected. This is why some of the new readings of Marx tend to get lost in the formal concordance of the history of the verbal expression of the theory of value in Marx's writings and fail to see the internal consistency of the theory as a whole. Moreover, a critique based on supposed theoretical inconsistencies derived from literal quotations can be confronted with a counter-critique based on other literal quotations that do not show such inconsistencies. This has been the case with the interpretation of Marx's Capital and his labor theory of value (Jones, 2016; Kliman et al., 2013). Even if Marx had renounced the law of the diminishing rate of profit, as has been speculated without any textual evidence, what matters is whether the law is consistent in itself and in the context of the whole theory of value, and whether it adequately reflects observed reality. In this respect, the empirical data on the long-term decline of the rate of profit is more than robust.

For these reasons, it is rather irrelevant whether Marx himself sometimes had doubts about the law of the rate of profit, or whether he tried and failed to find a more precise mathematical formulation or a clearer verbal exposition. After Marx, other researchers can take up the task that Marx could not undertake for a number of reasons, including his own limitations, the insufficient development of differential calculus or matrix algebra at his time, and other peculiarities of his personal life. This is how science proceeds, and this is how Marx's theory of labor is today a robust theory in which the periodic economic crises of capital accumulation, the decreasing rate of profit, and the habitability crisis are internally articulated showing consistent mutual determinations. Understanding capitalist production as a temporal single system in the light of the contemporary physics of complex systems, or unfolding Marx's concepts of social metabolism and metabolic rift to account for the current planetary crisis in the context of the labor theory of value are examples of positive developments of Marx's theory of value (Foster et al., 2010; Freeman &

Carchedi, 1996; Kliman, 2007). Splitting the laws or principles from the core of the theory on the basis of presumed inconsistencies in the history of the verbal expression of the theory and of speculations about Marx's intentions based on mixing such inconsistencies with Marx's personal life is a deconstruction of the theory not supported on scientific grounds.

Michael Heinrich is one of the main proponents of the *new reading* of Marx. According to Heinrich, the *Grundisse, A Contribution to the critique of political economy* and the *Manuscript of 1861–1863* correspond to a six-book project on the critique of political economy, while *Capital* is a different four-book project that "corresponds to the material of the first three books of the earlier six-book plan, but within an altered theoretical framework" (Heinrich, 2013a). From a comprehensive understanding of Marx's theory of value there is only one research project and only one theoretical framework, and the different versions of this project reveal the process of conceptualizing the theory. This is the usual procedure in any research and for any researcher, and so it is for Marx, whose theoretical framework is always the labor theory of value of the classical political economists, which he criticizes and improves on the basis of a dialectical and materialist epistemology. Marx's drafts show the history of the improvement of his method of research for a correct understanding of capitalist production, a method that finds its more complete version in Capital.[12] Cambridge University Library hosts the Charles Darwin Papers project, a collection of nearly all of Darwin's working scientific papers. No one has claimed to dismantle Darwin's theory of evolution by natural selection on the basis of the historical process of conceptualization of the theory, as shown in Darwin's Papers. Criticisms of Darwin's theory have been made on the basis of the theory itself, and they have actually improved it rather than dismantled it. For example, Engels' early critique of the struggle for life as the main mechanism of natural selection: "The whole Darwinian theory of the struggle for existence is simply the transference from society to organic nature of Hobbes' theory of bellum omnium contra omnes and of the bourgeois economic theory of competition, as well as the Malthusian theory of population" (Engels, 1986,

---

[12] The long-standing confusion in western Marxism about the exposition method and the research method in Capital has probably its origins in Rubin, who considers that Marx's exposition in Capital differs from the method of research he uses, see Rubin (1972). For one of the best understandings of Capital in terms of epistemology, see Ilyenkov (1982).

p. 181). Much later, Lynn Margulis and other researchers in the field of evolutionary biology have shown that symbiosis and cooperation played an important role in evolution by natural selection. Therefore, the question remains as to why Marx deserves such different treatment from other researchers with regard to the process of elaboration of his labor theory of value. In fact, Heinrich's exegesis of Marx's writings is based on a somewhat rigid interpretation of the texts and their historical sequence and prevents us from seeing not only the internal consistency of Marx's theoretical development but the development itself. This is because Heinrich's division of Marx's theoretical project into discrete segments is antithetical to a dialectical understanding of Marx's theory, which reveals the unity and evolution of the theory. As a result, Heinrich sees inconsistencies rather than the process of conceptualizing of a consistent theory and is forced to speculate about Marx's intentions based on such presumed inconsistencies:

> "Presumably, Marx was plagued by considerable doubts concerning the law of the rate of profit ... These doubts were probably amplified in the course of the 1870s ... In the case of a renewed composition of book III, all of these considerations would have had to find their way into a revision of the chapter on the "Law of the Tendency of the Rate of Profit to Fall." A consistent regard for them should have led to the abandonment of the "law" (Heinrich, 2013a).

Darwin's theory has been revisited, revised, and improved by many other researchers after Darwin, plate tectonics is based in part on the development and extension of Alfred Wegener's continental drift, Marx's theory of value is a critical revision of the classical political economy, and after him other researchers have undertaken a systematic treatment of capitalist crises on the basis of his theory of value. Heinrich blames Engels' editing of Capital, volume III for suggesting that "a systematic treatment of crisis theory" is "possible on the immediate basis of the law of the tendential fall in the rate of profit" and that "he created—on the part of readers who did not know that this chapter title did not at all originate with Marx—the expectation that this theory of crisis was a *consequence* of the "law" (Heinrich, 2013a, emphasis in the original). However, many readers who have properly understood the articulation of the law of the decreasing rate of profit and the phenomena of recurrent capitalist crises—likely without knowing that the chapter title is by Engels—do not pretend that there is a direct or immediate relationship between crises and the law, much less that the theory of crises in Capital is a *consequence* of the law. This is a mechanistic interpretation that cannot be attributed either to Marx or Engels (Mateo, 2018). Quite the contrary, the relation between the law and crises should be understood as a mediated relation between phenomena and essence, a relation in which the law is the structural and long-term trend of the capitalist production upon which singular crises of capital valorization unfold. This is Marx's understanding, for whom "all science would be superfluous if the outward appearance and the essence of things directly coincided" (Marx, 1967a, p. 592). In fact, Marx's conceptual development carried out mainly in Capital intertwines crises with the downward trend of profit, with capitalist accumulation, and with capitalist population dynamics within the theoretical framework of his labor theory of value, where all categories and laws are concretely interwoven in terms of causality and necessity and where the external forms of economic phenomena are related to their underlying essence.[13]

Heinrich's main objection to the diminishing rate of profit as such, i.e., not based on historical and philological exegesis of Marx's writings, is the alleged indeterminacy of the law.[14] Heinrich draws an analogy between Newton's law of gravity and Marx's law of the rate of profit to illustrate the indeterminacy of the latter, while the first can be precisely determined: "I already mentioned the difference between the quantitatively determined law of gravity and Marx's law which is not determined in the same way" (Heinrich, 2013b). This is strictly true. Marx's law cannot be determined *in the same way* as the law of gravity, that is on a mechanistic or Newtonian basis, but it can be determined, and the reason is that Marx's theory of value and its laws are closer to the physics of complex systems than to the simpler Newtonian physics. In fact, Marx's theory of value is said to describe a stochastic temporal system in which variables are empirically determined by means of probability density functions (Zachariah, 2009). Thus, the average profit and the prices formed on this average profit in the various economic sectors of capitalist production can

---

[13] For the dialectical relation between phenomena and essence, see Rosental and Straks, (1960), and Rosental (1962).

[14] Indeed, this objection is not new. The indeterminacy of the rate of profit was already formulated by Paul Baran and Joan Robinson, among others, in the first half of the twentieth century (Gill 2002, p. 509–511).

only be determined as gamma-type probability density functions (Cockshott & Cottrell, 1997; Kliman, 2012; Zachariah, 2009). Other economic variables, like the general profit of capitalist production and the composition of capital for the various economic sectors, cannot be determined at a given time on a mechanistic basis, but on a probabilistic one. Many natural and social processes are described as dynamic complex systems in which state variables cannot be determined on a mechanistic basis, but on a probabilistic basis. Examples include the forecasting of volcanic eruptions, the movement of molecules in a gas, the differentiation of species by natural selection from variations in the parent species, the movement of electrons around the atomic nucleus, and the dynamics of the world population. The question arises as to why a mode of production that has been established spontaneously, behind the back of individuals and through a historical process, should be different from natural systems that have no conscious predetermination, and the answer is that it should not. The temporal evolution of the variables of complex systems are usually modeled with Bayesian probabilistic and power law functions, as is the case of the gross domestic product, the surplus product and the world population (Golosovsky, 2010; Johansen & Sornette, 2001; Korotayev et al., 2016). These power law functions have intrinsic singularities defined as the time where the variable considered is infinite, which determines the impossibility of the system under study and a transition regime to another system. Of course, Marx could not have had an understanding on complex systems as it is known today. Nor could he, no matter how much interest he had in mathematics and other positive sciences, as is well known, have had an understanding of contemporary probabilistic, while such knowledge definitely exists today. That is why Marx intuitively understood the temporal evolution of some variables of the capitalist system as tendencies that have been later confirmed to be determined by probabilistic density functions and power law functions.

The rate of profit can be expressed in terms of surplus value and the composition of capital as $p' = s/v / (c/v) + 1$. Proponents of the empirical indeterminacy of the rate of profit consider the rate of surplus value and the composition of capital as independent variables that move freely and without limit in the course of capitalist production. In fact, on the surface of capitalist economic phenomena, it is observed that both variables, as well as labor productivity, grow without any apparent limit. In other words, there is a real factual basis for these considerations. Thus, by operating with the expression of the rate of profit in terms of money value and considering the variables independently, surplus value and its rate can grow without limit, as well as the value of constant capital and the composition of capital. In this way, the increase in the rate of surplus value can offset the increase in the composition of capital, that is, the expulsion of living labor from the immediate process of production. If the variables under consideration are independent and move freely, there is no way to characterize a priori the relationship between the respective rates, and so the rate of profit remains indeterminate. Implicit in this reasoning is that capitalist production has no intrinsic limits, that it has an absolute and infinite immanence, and its eventual finiteness, if any, will have an origin extrinsic to the capitalist mode of production. This is the opposite of Marx's understanding of capitalism as a historical mode of production, afflicted by insurmountable internal contradictions that ultimately determine its dissolution, and it is also opposite to the singularities described in power law functions of the capitalist system. Implicit in this view is that Marx's Capital, as a systematic unfolding of the contradictions of the reproduction of capital, is useless. The presumed empirical indeterminacy of the rate of profit eliminates the causal relationship between the variables involved in the expression of the rate of profit and, indeed, eliminates the fundamental internal contradiction of capital accumulation between capital and labor. The treatment of the variables as independent and in terms of money value rather than in terms of labor value, or at least not in those labor value terms that imply a complete unfolding of the material implications for the production process, results in an incomplete materialistic view of the rate of profit. On the contrary, if the rate of profit is considered in terms of labor value with all its material implications, and if the articulation of this law with other capitalist laws within the framework of the labor theory of value is also considered, the material determinations and limits of the mode of production are revealed. This is the materialist approach followed by Marx, as shown by his category of the organic composition of capital, which is determined by the technical composition of capital, and not the other way around: "To express this, I call the value composition of capital, in so far as it is determined by its technical composition and mirrors the changes of the latter, the *organic composition* of capital" (Marx, 1967b, p. 328 emphasis in the original.)

A materialist understanding of the rate of profit must also consider the limit to the increase in the rate of surplus value imposed by the transformation of necessary labor time into surplus labor time, which must be balanced in such a way as to allow the reproduction of the working class as a laboring class.[15] Capital seeks to alleviate this limit by increasing the number of workers, i.e. the population of the working class. However, based on Marx's law of capitalist population, the overpopulation of the working class determines the world's overpopulation, which has a limit set by the Earth's carrying capacity to support an increasing number of people living under capitalist conditions. On the side of constant capital, there is the limit imposed by nature's capacity to support an increasing exploitation of natural resources under capitalist conditions. Based on Marx's laws of capitalist accumulation, of capitalist population, and of the tendency of the rate of profit to fall, the working class accumulates as a necessity of the accumulation of capital and of the limit imposed on the rate of surplus value by the transformation of necessary labor time into surplus labor time. In this sense, the overpopulation of the working class can be seen as a countertendency to the fall of the rate of profit, which, together with the other countertendencies raised by capital to mitigate this fall, constitute empirical evidence of the fall itself.

In the historical course of social production, the social being is permanently assisted by working tools, which humans themselves develop and allow to increase the productivity of labor. Thus, the productivity of labor is qualitatively and quantitatively determined by the tools that support the production process, and both the number of tools used and the productivity of labor tend to increase in the course of history. This is a universal and general principle of social production, which also applies to the capitalist mode of production, in which the working tools have the historical form of constant capital. Capitalists produce only for profit, and since in the short term and in relation to rival capitalists, the more living labor is substituted by dead labor, the more profit is obtained, this substitution appears as the main determination of capitalist production. This general determination includes also the modern system of science and technology, which develops simultaneously with capitalism as a necessity of the reproduction of capital and has a feedback effect on the substitution of living labor, because it allows its increase. For these reasons, there is an accelerated increase in the constant capital involved in capitalist production processes that has no equivalent in human history. Therefore, the rate of surplus value and the composition of capital are not independent variables that move freely. Rather, the movement of the rate of surplus value

---

[15] See Sect. 3.3.1 *Declining rate of profit, economic crises and capitalist systemic crisis*.

is qualitatively and quantitatively constrained by the movement of the composition of constant capital, and not vice versa.[16] Thus, labor productivity, including productivity in those sectors that produce the means of subsistence of the working class, is determined by the composition of capital, not the other way around. The logical conclusion is that the rate of surplus value can only occasionally grow faster than the composition of capital, but not in the long term of the capitalist mode of social production. This is why, on the basis of Marx's dialectic understanding, the great unfolding of productive forces in capitalism is seen as a paradox, as the erosion of the motive power of capitalist production, that is, the rate of profit.

3.3.2.2 David Harvey's Critique of the Profit Rate and Crises Theory

David Harvey's concept of capitalist crises is closer to that of Keynes than to that of Marx. According to Keynes, capitalist crises are crises of effective demand that respond to multiple contingencies and are "the mere consequence of upsetting the delicate balance of spontaneous optimism. In estimating the prospects of investments, we must have regard, therefore, to the nerves and hysteria and even the digestions and reactions to the weather of those upon whose spontaneous activity it largely depends" (Ilyenkov, 2012).[17] According to Harvey, crises are multi-causal processes with such complexity that it is not possible to understand them at the core of the organic theory of value. Like Keynes, Harvey believes that the multiple contingencies and subjectivities involved in capitalist crises make impossible to articulate them theoretically around some of the basic principles of capitalist production. In other words, it is not possible to characterize the structural and objective determinations of crises in terms of causality and necessity within the core of Marx's theory of value and in relation to other laws of this theory.

According to Harvey, the multiple and particular vicissitudes of each crisis should be studied individually, and from the multiple phenomena

---

[16] The quantitative dependence of the rate of surplus value on the composition of capital in the equation of the profit rate has been shown with partial derivatives by Gill (2002).

[17] Ilyenkov provides this quotation from Keynes to illustrate his metaphysical understanding of the relationship between the material and the ideal, which transforms into subjective idealism and underlies Keynes' conception of the economic relations.

observed in many crises, a general description should be deduced with the general features common to many crises. This is certainly a necessary step for any scientific understanding of reality, but it is also insufficient. The inductive abstraction of the common features observed in phenomena is only one moment in the process of scientific understanding of reality, in order to classify the multiple phenomena perceived by the senses. It allows the subsequent unfolding of the theoretical concepts that explain the observed phenomena not only descriptively, but also relate them in terms of their causal concatenations and necessities. Logical inference by deduction plays a major role in this process and completes the inductive abstraction.[18] In this way, the laws or general principles that describe the essentials of the system under study can be characterized, and the particular phenomena can be understood as mediated expressions of these laws. Without this conceptual development and the logical-deductive analysis of the connections between concepts, the theory remains incomplete, or rather, it is not a theory but a description of phenomena that does not allow for predictive interaction with the system. Inductive reasoning based on empiricism and deductive reasoning are two opposite but inseparable moments of the process of ascending from the abstract to the concrete, which is "the specific form of the thinking activity, of the logical elaboration of contemplation and representation in concepts" and which allows for the scientific understanding of the multiple determinations synthesized in the concrete reality (Ilyenkov, 1982).

Classical political economy culminated its theoretical development with David Ricardo, who already had the concept of abstract social labor as the substance of value. However, Ricardo failed to go further in his theoretical development when trying to find a direct, unmediated manifestation of abstract social labor in particular forms of value (profit, wage, rent, etc.). Ricardo's approach proceeds by formally abstracting the common features of the various forms of value and then trying to deduce the expression of abstract social labor in these forms, but without studying the concrete mediations involved (Ilyenkov, 1982). With this methodological and epistemological approach, Ricardo failed to understand the concrete relationship between abstract social labor and the forms of value, a failure that has not been remedied by post-Ricardo bourgeois economics. Marx, with the full range of categories provided by classical

---

[18] For the dialectical unity of induction and deduction and its role as a mechanism of the inferred reasoning, see Rosental (1962).

political economy and the invaluable legacy of Hegel's logic, was able to identify the form of value that is fundamental to the system under study. Abstracting this form from all other forms of value, he analyzed the simplest, most concrete, and universal relationship that appears on the surface of the economic phenomenon: the exchange between two commodities. This analysis is carried out on a rigorously dialectical and materialist basis, in which dialectical logic and the unfolding of contradictions in categories play a leading role, which does not exclude formal logic as a necessary moment of theoretical elaboration. Broadly speaking, this is Marx's methodological and epistemological procedure in Capital.

In this way, Marx is able to unfold the logical succession of the forms of value, which roughly coincides with their historical unfolding, and the concrete mediations that connect them, and to formulate the general laws of the system from which these forms can be understood in relation to the essentials of the system.[19] In Marx's theory of value, the forms of value unfold from the fundamental contradiction of the commodity form, that is, from the contradiction between use value and value, which is expressed in the exchange of commodities as the contradiction between use value and exchange value. Since this is a very real contradiction, all forms of value carry it in their core, albeit manifested through concrete mediations. For example, this fundamental contradiction is expressed as a decline in the rate of profit when it is mediated by profit and the competence for profit, or the prices of production are the forms of value mediated by the average rate of profit and express the fundamental contradiction as a price disequilibrium. In fact, the unfolding of the fundamental contradiction and its mediated expressions is the driving mechanism for the historical development of capitalism, and the logical contradictions with which Marx works are only a reflection of these real contradictions and their historical unfolding. According to this view, the falling rate of profit, periodic economic crises, the capitalist systemic crisis, and the planetary crisis are all mediated expressions derived from the fundamental contradiction of the capitalist mode of social reproduction.

Marx and Harvey hold antagonistic views regarding economic crises, the crisis theory, and other significant elements of the labor theory of value. For Marx, the depauperation of the industrial reserve army and

---

[19] On the relation between the historical and the logical in the process of understanding, see Rosental and Straks (1960)

the decreasing rate of profit are constitutive laws of capitalist production, while for Harvey they are contingencies:

> "Marx's theory of the falling rate of profit should be treated as a contingent rather than a definitive proposition. It says, in effect, that if there is a fall in the rate of profit here is one of many ways in which it could come about. ... Whether or not this particular mechanism is the one at work depends, however, upon careful analysis of actually existing dynamics. My own guess is that crises produced by this mechanism are relatively rare" (Harvey, 2016, p. 11).

Harvey understands the relationship between the rate of profit and economic crises in terms of direct cause-and-effect processes, in which the declining rate of profit is a contingent mechanism that rarely causes crises. But understanding the articulation of the rate of profit and crises in terms of causality and necessity, as Marx does, certainly does not mean such a direct and immediate relationship. Rather, it means distinguishing the immanent or ultimate determinations of the system from the contingencies manifested in phenomena and investigating the concrete mediations that connect them in terms of causality and necessity. That is, it means investigating the structural laws of the system and the particular circumstances under which these laws are expressed in phenomena. In this way, a comprehensive understanding of the complexity of the system is possible, which is otherwise not possible in terms of direct cause-and-effect relationships. Marx's methodological and epistemological approach makes it possible to integrate the various economic phenomena of the capitalist system, such as the rate of profit, periodic crises, financial hypertrophy, and even the need to print money unsupported by value and the erratic policies of central banks, which can certainly trigger a given particular crisis, into the consistent organic corpus of the labor theory of value.

The epistemological approach followed by Marx allows him to understand the tendency of the rate of profit to fall as a limit of capitalist production "In that the development of the productivity of labour creates out of the falling rate of profit a law which at a certain point comes into antagonistic conflict with this development and must be overcome constantly through crises" (Marx, 1967a, p. 178). A very crude analogy with plate tectonics can be drawn to illustrate how an epistemological procedure similar to Marx's works in the natural sciences. Lithospheric plates move at velocities of a few millimeters per year. At some plate boundaries, stress is concentrated, depending on a number of factors such as plate boundary topology, local fluid pressure, and others. Such stress can be suddenly released in earthquakes, eventually killing thousands of people. Earth scientists do not claim a direct cause-and-effect relationship between plate motion and earthquakes without considering more local and contingent features, nor do they claim that fluid pressure caused the earthquake without considering long-term and structural plate motion. On the contrary, they proceed by integrating all available data at different scales and hierarchies to build a consistent theoretical corpus that can account for the observed complexity of the phenomena. Based on these epistemological grounds, it is possible to reconstruct past plate dynamics and to predict future plate organization, although neither can ever be an exact copy of reality, only an approximation. In this respect, Marx's understanding of the falling rate of profit as a long-term insurmountable limit of capitalist production can be seen as a kind of primitive forward modeling of a historical mode of production. The question arises as to why a mode of production like capitalism, with blind dynamics that escapes humans' control, should not have structural principles or laws similar to those of the natural systems, and the answer is that it should, but as far as the social sciences are concerned, the dominant idealist view is not inclined to accept that the social and economic system built on the dominance of the bourgeois class is governed by laws that the class itself does not understand and does not control. On the contrary, the materialist and dialectic view of Marx allows for deciphering "the economic law of motion of modern society," which is "the ultimate aim" of Capital (Marx, 1967b, p. 7).

3.3.3 *Habitability Crisis, Capitalist Systemic Crisis, and Decreasing Rate of Profit*

For Marx, it is clear that material reality is the origin of sensory perception and the subsequent conceptual elaboration through thought, where this reality is reflected and a mental image is obtained, and that theoretical conceptualization is returned to material reality through practical activity, which ultimately sanctions whether the reflection in thought is adequate to reality itself or not. Marx develops the concept of metabolic rupture as a necessity of the overproduction and accumulation of capital, which breaks the nutrient cycle of human social metabolism through the increasing plundering of nature and the returning of an increasing toxicity to nature. The accumulation of capital in cities, industrial poles, and industrial agriculture accelerates the metabolic rupture with respect to the social metabolism of previous modes of production. Marx could not have the concept of the crisis of habitability in its full dimension, because such a crisis was not a dominant material reality in his time. However, Marx's concept of the metabolic rift is a logical deduction from the overaccumulation of capital at a decreasing rate of profit, and thus it can be read as a precursor of the current crisis of habitability. When the analysis of the reproduction of capital in terms of value is transposed to the material reality that ultimately determines this reproduction and when the environmental implications of this analysis are derived for a world that is by definition finite, the structural link between the systemic capitalist crisis and the current habitability crisis is revealed. Such an analysis must be carried out in the context of the labor theory of value, taking into account all the elements and principles of the theory in which the tendency of the rate of profit to fall plays a fundamental role.

Capital must accumulate in order to reproduce, and accumulation at a diminishing rate of profit forces the various capitals to overproduce commodities in order to increase their share of the total profit produced in society, which they share competitively. The overproduction of commodities necessarily entails, on the one hand, the overexploitation of the natural resources that feed the productive processes and, on the other hand, the productive overconsumption of these resources and the consequent generation of waste materials whose environmental toxicity depends not only on their physical and chemical properties but also on their quantity. Moreover, commodities must be individually consumed so that the reproductive cycle of capital can be completed, an individual consumption that in turn generates new waste materials of varying toxicity. The more capital is forced to overproduce, the more waste materials are generated, and the more capital overproduces with a decreasing rate of profit, the more it must accumulate with an increasing organic composition and the more it must concentrate and centralize. To paraphrase Marx, the accumulation of capital at one pole is the accumulation of the working class that produces that capital with its labor at the opposite pole, and it is also the accumulation of environmental toxicity on the Earth.

A declining rate of profit drives and accelerates the process of capital accumulation in all directions. It is a pumping mechanism, without analogy in the history of the Earth and of humanity, by which wastes of various toxicity accumulate in a nonlinear and accelerated way. That is why it is so dangerous. In essence, this can be read as the general law of environmental degradation under capitalism. Accumulation, concentration, and centralization of capital imply the development of large cities and industrial poles where fixed capital is accumulated and where circulating capital is processed by variable capital. With the historical unfolding of capitalist production on Earth, this implies the exchange of commodities on a planetary scale as a necessary moment of the production process, which in turn adds new waste materials to be absorbed by the Earth's ecosystem. This capitalist dynamic entails the physical occupation of planetary space, eroding those natural spaces and spheres of human life where the capitalist roller has not yet arrived.

Such an expansive dynamic is fueled by a decreasing rate of profit and leads to the conclusion that the origin and development of the crisis of habitability has an economic basis and that the more the rate of profit decreases, the worse the habitability crisis will be. Capitalist humans engaged in the blind dynamics of capital accumulation at a decreasing rate of profit are not aware of what they do, but nevertheless, they do it. As shown by empirical data on the rate of profit, after about a century of capital accumulating at a decreasing rate of profit, the crisis of habitability has become such a self-evident reality that capitalist humans have finally become aware of it. Thus, the secular crisis of valorization of value riding on the back of the falling rate of profit has resulted in a crisis of habitability that is immanent to the reproduction of capital. For this reason, if the systemic crisis reveals the internal limit of capitalist production, and its potential historical finitude, the crisis of habitability reveals the potential absolute limit not only of the capitalist mode but of all forms of social reproduction.

In their social metabolism with nature, human beings must necessarily submit to the laws of nature, which are the ultimate determination, the possibility, of any form of social metabolism. Knowledge of these laws is a prerequisite for a conscious social metabolism with nature, and on the basis of this knowledge, it is possible to conceive a human fit with the planet in accordance with nature. However, an adequate fit with nature is not possible under the capitalist mode of production, because the reproduction of capital imposes its own laws and subordinates the laws of nature to its own metabolism. In other words, in the capitalist mode, the natural laws are subsumed under the laws of the reproduction of capital. For this reason, the capitalist form of human social metabolism with nature is necessarily an alienated form beyond human control. Natural laws are universal, they are constitutive of nature and all forms of social organization, while capitalist laws are historical, they are universal only for a particular mode of production. Hence, both nature and humans have the possibility to change the capitalist laws. Nature because it has the last word on everything, and humans because they can change the mode of production on the basis of the knowledge and ethics intrinsic to the teleological character of labor, human's living practical activity. It is a paradox that the enormous scientific knowledge of nature developed under capitalism has led to a planetary crisis without analogy in the history of the Earth, a paradox that has no possible solution within the limits of capital reproduction. The great unfolding of productive forces under capitalism based on scientific knowledge has led to both a real and at the same time apparent decoupling of humans from nature, and this is one of the main obstacles to transcending the capitalist mode of social production because it is idealized as absolute on the basis of the power it unfolds.

References

Basu, D., Huato, J., Lara, J., Wasner, E. (2022). *World profit rates, 1960–2019*. Economics Department Working Paper Series 318. https://doi.org/10.7275/43yv-c721

Callinicos, A., & Choonara, J. (2016). How not to write about the rate of profit. *Science & Society, 80*, 481–494.

Carchedi, G., & Roberts, M. (Eds.). (2018). *A global analysis of Marx's law of profitability*. Haymarket.

Cockshott, P., Cottrell, A. (1997). *The scientific status of the labour theory of value*. Available at: https://users.wfu.edu/cottrell/eea97.pdf

Cockshott, P., Nieto, M. (2017). *Cibercomunismo. Planificación económica, computadoras y democracia*. Editorial Trotta

Cunha, D. (2015). The geology of the ruling class? *The Anthropocene Review, 2*, 262–266.

Del Rosal, M. (2019). *La gran revelación: De cómo la Teoría Monetaria «Moderna» pretende salvarnos del capitalismo salvando el capitalismo*. Ecobook-Editorial del Economista.

Ehrlich, P. R., & Ehrlich, A. H. (2012). Solving the human predicament. *International Journal of Environmental Studies, 69*, 557–565.

Ellis, E., Maslin, M., Boivin, N., & Bauer, A. (2016). Involve social scientists in defining the Anthropocene. *Nature, 540*, 192–193.

Engels, F. (1947). *Anti-Dühring. Herr Eugen Dühring's Revolution in Science.* Progress Publishers.
Engels, F. (1986). *Dialectics of nature.* Progress Publishers.
Fineschi, R. (2008). *Un nuovo Marx. Filologia e interpretazione dopo la nuova edizione storico critica (MEGA2).* Caroci.
Fischer-Kowalski, M., Krausmann, F., & Irene Pallua, I. (2014). A sociometabolic reading of the Anthropocene: Modes of subsistence, population size, and human impact on Earth. *The Anthropocene Review, 1,* 6–31.
Foster, J. B. (2013). Marx and the rift in the universal metabolism of nature. *Monthly Review, 65,* 1.
Foster, J. B., & Burkett, P. (2018). Value isn't everything. *Monthly Review, 70,* 1–17.
Foster, J. B. (2017). Marxism in the Anthropocene: Dialectical rifts on the left. *International Critical Thought, 6,* 393–421.
Foster, J. B., Clark, B., & York, R. (2010). *The ecological rift: Capitalism's war on earth.* Monthly Review Press.
Freeman, A. (2019). *The sixty-year downward trend of economic growth in the industrialised countries of the world.* GERG Data Group working paper No.1. Geopolitical Economy Research Group
Freeman, A. (2013). The profit rate in the presence of financial markets: A necessary correction. *Journal of Australian Political Economy, 70,* 167–192.
Freeman, A., & Carchedi, G. (Eds.). (1996). *Marx and non-equilibrium economics.* Edward Elgar.
Gandler, S. (2006). Releer a Marx en el siglo XXI. Fetichismo, cosificación y apariencia objetiva. *Dialéctica, 38,* 119–134.
Gill, L. (2002). *Fundamentos y límites del capitalismo.* Editorial Trotta.
Golosovsky, M. (2010). Hyperbolic growth of the human population of the Earth: Analysis of existing models. In L. Grinin, P. Herrman, & A. Korotayev (Eds.), *History and mathematics: Processes and models of global dynamics* (pp. 188–204). Uchitel.
Guillaume, B. (2014). Vernadsky's philosophical legacy: A perspective from the Anthropocene. *The Anthropocene Review, 1,* 137–146.
Hamilton, C. (2016). The Anthropocene as rupture. *The Anthropocene Review, 3,* 93–106.
Harvey, D. (2016). Draft of Crisis theory and the falling rate of profit. In: T. Subasat, (Ed.) *The Great Financial Meltdown.* Edward Elgar, Cheltenham. Available at: https://thenextrecession.wordpress.com/wp-content/uploads/2014/12/harvey-on-ltrpf.pdf
Heinrich, M. (2013a). Crisis theory, the law of the tendency of the profit rate to fall, and Marx's studies in the 1870s. *Monthly Review, 64,* 15.
Heinrich, M. (2013b). Heinrich answers critics. *Monthly Review, 64,* 15.

Hornborg, A. (2013). *Global ecology and unequal exchange: Fetishim in a zero-sum world*. Routledge.
Hornborg, A., & Malm, A. (2016). Yes, it is all about fetishism: A response to Daniel Cunha. *The Anthropocene Review, 3*, 205–207.
Ilyenkov, E. V. (1982). *Dialectics of the abstract and the concrete in Marx's capital*. Progress Publishers.
Ilyenkov, E. (2012). Dialectics of the Ideal. *Historical Materialism, 20*, 149–193.
Johansen, A., & Sornette, D. (2001). Finite-time singularity in the dynamics of the world population, economic, and financial indices. *Physica A: Statistical Mechanics and Its Applications, 294*, 465–502.
Jones, P. (2016). Devaluation and Marx's Law of the tendential fall in the rate of profit. *Review of Political Economy, 28*, 233–250.
Kapitza, S. (1996). The phenomenological theory of world population growth. *Physics-Uspekhi, 39*, 57–71.
Kliman, A., Freeman, A., Potts, N., et al. (2013). *The unmaking of Marx's capital: Heinrich's attempt to eliminate Marx's crisis theory*. Available at: http://papers.ssrn.com/sol3/papers.cfm?abstract_id=2294134
Kliman, A. (2007). *Reclaiming Marx's capital: A refutation of the myth of inconsistency*. Lexington Books.
Kliman, A. (2012). *The failure of capitalist production*. Pluto Press.
Korotayev, A. V., & Malkov, A. Y. (2016). A compact mathematical model of the World System economic and demographic growth. *International Journal of Mathematical Models and Methods in Applied Sciences, 10*, 200–209.
Kremer, M. (1993). Population growth and technological change: One Million B.C. to 1990. *The Quarterly Journal of Economics, 108*, 681–716.
Krugman, P. (2012). *Rise of robots*. The New York Times
Maito, E. (2016). El capitalismo y su tendencia al derrumbe. *Revista En Defensa Del Marxismo, 48*, 125–171.
Martínez Marzoa, F. (1981). *La filosofía de El Capital de Marx*. Taurus.
Marx, K. (1959). *Economic & philosophic manuscripts of 1844*. Progress Publishers.
Marx, K. (1967a). *Capital 3*. International Publishers.
Marx, K. (1967b). *Capital 1*. International Publishers.
Marx, K., & Engels, F. (1976). *The German ideology*. Progress Publishers.
Mateo, J. B. (2018). Marx's law of the profit rate and the reproduction of capitalism: Neither determinism nor overdetermination. *World Review of Political Economy, 9*, 41–60.
Mészáros, I. (2010). *Social structure and forms of consciousness*. Monthly Review Press.
Oldfield, F. (2018). A personal review of the book reviews. *The Anthropocene Review, 5*, 97–101.

Palsson, G., Szerszynski, B., Sörlin, S., et al. (2013). Reconceptualizing the 'Anthropos' in the Anthropocene: Integrating social sciences and humanities in global environmental change research. *Environmental Science & Policy, 28*, 3–13.

Piedra Arencibia, R. (2019). *Marxismo y dialéctica de la naturaleza*. Edithor.

Pigmaier, E. (2021). The value of value theory for ecological economics. *Ecological Economics, 179*, 106790.

Rosental, M., & Straks, G. M. (Eds.). (1960). *Categorías del materialismo dialéctico*. Editorial Grijalbo.

Rosental, M. (1962). *Principios de lógica dialéctica*. Ediciones Pueblos Unidos.

Rubin, I. I. (1972). *Essays on the Marx's theory of value*. Black and Red.

Smith, M. E. G., Butovsky, J., & Watterton, J. (2021). *Twilight capitalism: Karl Marx and the decay of the profit system*. Fernwood Publishing.

Swanson, H. A. (2016). Anthropocene as political geology: Current debates over how to tell time. *Science as Culture, 25*, 157–163.

Van Noorden, R. (2013). Who is the best scientist of them all? *Nature News*. Available at: https://www.nature.com/articles/nature.2013.14108

Zachariah, D. (2009). Determinants of the average rate of profit and the trajectories of capitalist economies. *Bulletin of Political Economy, 3*, 1–13.

Zeller, T. (2015). *You are an epoch: Defining the Anthropocene*. Available at: http://www.forbes.com/sites/tomzeller/2015/03/11/you-are-an-epoch-defining-the-anthropocene/#54c745b94848

CHAPTER 4

Epistemological Grounds to Transcend the Habitability Crisis

Marx's Eleventh Thesis on Feuerbach states, "Philosophers have only *interpreted* the world, in various ways; the point is, however, to *change* it" (Engels, 1946, Appendix, emphasis in the original). The planetary crisis of habitability is certainly demanding a change of the world if the Earth is to remain habitable. Such a change is possible only on the basis of a scientific understanding of the world itself, that is, an understanding of the objectivity of the world, of the world as it is, and not on the basis of intuition, desire, or speculation, which can be of some help in the task of changing the world, but not enough. In the challenge of changing the world as posed by the crisis of habitability, some transcendental questions arise, such as what is science? what is nature? what is society? can human society be understood on a truly scientific basis? These are essentially philosophical, particularly epistemological, questions that the natural sciences themselves cannot answer. Despite the importance of philosophy and epistemology, however, they are rather neglected or misunderstood in current studies of the natural sciences.

Human beings are confronted with an objective reality, whether social or natural, which must be understood on the basis of its own objectivity in order to be transformed and to satisfy human needs. Such an understanding implies theoretical knowledge, without which a practical transformation of reality is not possible. This is the meaning of the scientific socialism of Marx and Engels, later followed by Lenin and most of the philosophers in the Soviet Union (Engels, 1970). That is, socialism has to be scientific, or not to be. Since the only theoretical corpus we have for the capitalist society is Marx's theory of value, which is still fully operative, this theory must be understood on a Marxian basis. In other words, not all interpretations of Marx' theory are possible if they do not reflect the reality of capitalist production and depart from Marx's epistemological approach, which has proven to be correct for the understanding of capitalism.

4.1 The Importance of Philosophy for Scientific Understanding

The social sciences usually assume the important role of philosophy in their studies, and to some extent, the social sciences and humanities are the natural habitat of philosophy. On the contrary, natural sciences tend to see philosophy as a foreign discipline unrelated to the object of their study, nonetheless, many of the disputes in the natural sciences about various issues throughout history have an epistemological basis in the background. The importance of philosophy and epistemology in resolving dilemmas in the natural sciences has historically been more evident in the fields of physics and biology.[1] For example, the crisis of modern physics at the beginning of the twentieth century, in which the possibility of motion without matter, and thus the disappearance of matter, was considered, and led some notable physicists like Poincaré to conclude that "it is not nature which imposes on [or dictates to] us the concepts of space and time, but we who impose them on nature"; "whatever is not thought, is pure nothing" (Lenin, 1974). Lenin corrected Poincaré for his idealistic view and, following Engels, provided a materialist and monist understanding of the dialectical unity of matter and motion. Later on, during the Bohr-Einstein debates about quantum mechanics, Einstein declared that.

---

[1] See Piedra Arencibia (2016) for an excellent contribution on the importance of philosophy in biology and physics, which highlights some misconceptions of leading contemporary scientists regarding the categories of determinism and freedom, which are ultimately related to a non-dialectical, i.e. metaphysical, understanding.

"he would have liked 'old Spinoza' as the umpire in his dispute with Niels Bohr on the fundamental problems of quantum mechanics rather than Carnap or Bertrand Russell, who were contending for the role of the 'philosopher of modern science' and spoke disdainfully of Spinoza's philosophy as an 'outmoded' point of view 'which neither science nor philosophy can nowadays accept'". (Ilyenkov, 1977)

The reason for Einstein's preference for Spinoza was his truly materialistic approach, as opposed to the idealistic and positivistic understanding of Carnap and Russell. More recently, the debate about the feasibility of the so-called Gaia hypothesis or theory has brought to the fore questions such as whether there is any kind of conscious or unconscious teleology in nature, and what are the limits of empiricism for validating theories (Bondi, 2015; Janković & Ćirković, 2016; Kirchner, 2003; Lenton, 1998). Other questions raised, such as what are the units of natural selection, whether individuals and taxa or metabolic and developmental interaction patterns resemble the matter versus motion debate in early twentieth-century physics and cannot be properly addressed from a dichotomous perspective, but only from a monistic understanding (Doolittle & Booth, 2017).[2]

These are just a few examples to illustrate how philosophy can provide the right focus to solve problems and dilemmas that the natural sciences usually face in the study of natural phenomena. Modern Earth system science and Anthropocene studies are no exception, and they cannot neglect philosophy in the face of the problems that human societies pose for the understanding of nature, including those related to the formalization of the Anthropocene. This is because the concepts and categories of the natural sciences, including modern Earth system science and Anthropocene studies, evolve throughout history according to the general evolution of the epistemological paradigms. These concepts and categories derive from and at the same time shape such epistemological paradigms. Thus, the formalization of the Anthropocene cannot dispense with the concept of the Anthropocene as a crisis of habitability and with a

---

[2] The references given here are just a few examples of the debate about the Gaia hypothesis.

given and historical conceptualization of the relationship between humans and nature.[3] As stated by Valery Bosenko, these are.

> "philosophical questions of physics, philosophical questions of biology, etc., in short, the philosophical questions of the natural sciences. These are solved by naturalists themselves (and not by philosophers) *with the help of* philosophy. For this, the naturalists wield (they must necessarily take up) the philosophical weapons, the materialistic principles, dialectical principles, mastering dialectical theoretical thinking (*creating it each one by himself*), dialectical logic, gnoseology, etc.". (Bosenko, 1965, emphasis in the original, English translation from Spanish by the author)

4.1.1 Science and Scientific Knowledge

Scientific knowledge seeks to understand the multiple and mutually interrelated determinations whose synthesis shapes the concrete totality of the system under study. This requires theoretical elaboration by thought of phenomena perceived by senses. That is, it requires the transformation of the empirical phenomena given to us by practice and experience into concrete concepts articulated within an organic theory capable of grasping the constitution of the studied object and its movement or history. Scientific understanding means knowing the specific role of each determinant in the configuration and origin of the system under study so that its evolution can be predicted and a practical interaction with the system can be carried out. From a methodological perspective, the abstraction of concrete concepts through theoretical thought is a remarkably different process than the abstraction of generalities from the observed phenomena. Essentially, the first is deductive while the second is inductive, and both are necessary and complementary moments for any understanding of the concrete reality. Empirical abstractions by induction are necessary in the process of theoretical thought and have the descriptive role upon which theoretical deduction can be based. Induction, deduction, analysis, synthesis, conceptual and judgment elaboration, and others are forms of thought and of the process of inferred reasoning that allow the ascent from the abstract to the concrete in the process of

---

[3] For an epistemological assessment of the formalization of the Anthropocene in the Geologic Time Scale and of the disputes about formalization, see Sects. 2.1.4 and 2.1.5, and Soriano (2024).

understanding reality. The ascent from the abstract to the concrete is the general law or principle of scientific understanding, actually of any understanding of reality.[4] Synthetically, this principle states that the starting point of knowledge is the material reality out of human thought, and that a process of conceptual abstraction of the phenomena perceived by the senses is necessary to reveal the essences, regularities, or laws underlying the observed phenomena. Once these laws have been established, further research makes it possible to return to phenomenal reality in the light of the concrete mediations that link the underlying essence to the concrete reality perceived. Abstract and concrete are thus two contradictory moments of the cognitive process, in which abstraction is a vanishing moment of the whole cognitive process and concrete reality is the starting point and the final point of understanding. From a gnoseological point of view, the concrete reality that emerges from the process of abstraction and the study of the concrete mediations that shape reality is different from the initial concrete phenomena perceived by the senses. Although it may seem paradoxical, the more concrete our understanding or reality is, the more abstract it must be. In summary, in the ascent from the abstract to the concrete, concrete reality is reproduced in thought.

Logic is the science that studies the forms and processes of thought, but not from a physiological perspective, which is the object of study of neuroscience, but from the perspective of the mechanisms of reasoning that human thought has necessarily to interpose in its relation to reality.[5] The ability to perform a material transformation of reality deciphers to what extent a scientific understanding of a given system is correct or not. Hence, practice is the ultimate goal, it is an end in itself beyond which no other end is possible. Abstraction of generalities by induction and theoretical elaboration by logical deduction are a means of dynamic interaction with reality, they are vanishing moments with respect to practice. However, whatever the material transformation of reality is, it always appears as an objective reality independent of the way in which it is

---

[4] For a detailed unfolding of this general principle of logic and epistemology, see Ilyenkov (1982a), Rosental (1962), Rosental and Straks (1960).

[5] Neurons and neural processes of the brain are only the physiological support upon which induction, deduction, concepts, judgments, and all forms of thought interact to make cognition possible. Without neurons and their activity, thought is impossible, but they do not constitute the process of thought as such. Here, verbal language is both the main vehicle and the form of expression of thought (Ilyenkov, 1982a; Rosental, 1962).

perceived and thought, and for this reason it can be scientifically known. In a broad sense, rational thought is understood "as the ideal component of the real activity of social people transforming both external nature and themselves by labor" (Ilyenkov, 1977, p. 2). Labor is here defined as the practical activity capable of transforming reality. It is the human universal par excellence, with an immanent teleological character, which implicitly requires the knowledge of ends and means and the adoption of ethical choices regarding the particular activity undertaken as well as the results obtained (Lukács, 1980). From an anthropological perspective, labor is the specific activity by which humans have evolved throughout history into human social being.

Scientific knowledge is here understood as a social product carried out by a collective social being throughout history, which is culturally inherited across history and in which individuals participate as active members. They, themselves, are determined by the structures of a given historical society, with a given historical scientific knowledge that they submit for critical examination over the course of their scientific research. A collective social being means that society is not an aggregate of individuals interacting with each other according to their particular interests, where the state or God is necessary and acts as abstract and supposedly independent regulators. This would be a kind of Hobbesian-Hegelian conception of society, which certainly corresponds to some historically determined types of society, but is by no means a human universal. Rather, society is conceived here as a social being in which individuals evolve collectively, in other words, a social being that evolves with and because of the evolution of individuals and vice versa. In this sense, science can be conceptualized as the never-ending activity of the social being aimed at understanding concrete reality and enabling its practical interaction and transformation. To paraphrase Samuel Beckett's famous quote, in scientific activity people "fail, fail again, fail better." It is beyond the scope of this section to provide a detailed analysis of human understanding throughout history, including religious forms, crude materialism and empiricism, ancient and Hegelian dialectics, positivism, postmodern relativism, and so on, which would certainly reveal a nonlinear progress of scientific knowledge. Nevertheless, it seems quite plausible to postulate a rough evolution of human knowledge from an abstract and idealistic understanding based on myths and gods to a more concrete and materialistic understanding of reality that allows a more practical and concrete interaction with it. Therefore, to deal with science is to deal with things as they are, it is to deal with the process of knowing a relatively absolute truth. In this regard, it is worth remembering Marx's claim on abandoning any sort of ideological prejudice about science:

> "At the entrance to science, as at the entrance to hell, the demand must be made:
>
> > Qui si convien lasciare ogni sospetto
> > Ogni vilta convien che qui sia morta" (Marx, 1993, Preface).[6]

4.1.2 Natural and Social Science Disciplines

The division of scientific knowledge into natural and social science disciplines is a formal division promoted by neo-Kantian philosophers in the nineteenth century, partly in response to Hegelian philosophy. Prior to this rupture, most philosophers and thinkers, including those of earlier modernity—Steno, Leibniz, Petty, Descartes, etc.—had a broader and more inclusive approach that did not make such a sharp distinction between the social and natural sciences. The most relevant formal difference between the fields of social and natural sciences is related to the time scale over which the respective objects of study evolve compared to the time scale over which theories about those objects evolve.

The object studied by the natural sciences usually does not change in the course of the historical development of its theoretical understanding. In other words, the time scale of the evolution of the studied object is on such orders of magnitude that it can be neglected with respect to the time scale on which the various theories about the object evolve. This is the case in physics, biology, geology, and most of the natural sciences. For example, the theoretical understanding of the cosmos has evolved significantly from geocentrism to heliocentrism to gravitational physics to relativity, but the object of study has remained the same throughout this evolution. Similarly, the theoretical understanding of the origin of life and the evolution of species has changed since Linnaeus, Darwin, Oparin, and up to the present genetic theory, but the object of study is the same for all of them. In these cases, old theories are generally simple and abstract expressions of theoretical understanding of the object, while newer theories are more complete and concrete, the whole theoretical

---

[6] Marx pharaphrases Dante's Inferno regarding the suitable attitude in front of science.

development corresponding to the *failing, failing again, failing better* process of understanding. For example, Alfred Wegener's continental drift is a simple and abstract theory with respect to the more complete and concrete plate tectonics.

The situation is different in the social sciences. Here, the evolution of the system under study occurs on roughly the same time scale as the evolution of the theories about it. In these cases, the historical development of theories roughly mirrors the historical development of the object of study. Moreover, the different theories of the system throughout history provide different forms of dynamic interaction with the system, which may interfere with its development. This may modify the phenomenal forms in which the object under study appears and, accordingly, the inductive abstraction of generalities inferred from empirical observation and sensory perception. For example, the theoretical understanding of social modes of production has evolved largely concurrently with human's practical interaction with nature, leading to its material transformation and the various modes of production observed throughout human history. Thus, Aristotle could not have a concrete concept of abstract labor because abstract labor as a social average did not exist in his time. In contrast, David Ricardo was able to understand abstract labor as the substance of value because abstract labor as a social average constitutes the social reality of capitalist production.

The formal differences between the natural and social sciences explain why naturalists can see themselves as external observers of the system under study, while social scientists cannot. Although a deeper understanding of the reality of natural systems usually reveals that the activity of scientific research interferes with the studied system, in most cases, and for practical purposes, such activity can be neglected. For example, if we consider the measurement of the temperature of the ocean, it implies a mutual interaction, a change in both the temperature of the ocean water and the temperature of the thermometer (Piedra Arencibia, 2016). In these cases, as in the scientific study of the history of the earth, the history of life, and the history of the cosmos, humans can neglect the interference of their activity on the objects they study, although such interference obviously exists. However, when dealing with social systems, an approach that places us as external observers of the system is no longer possible because practical interaction with the system strongly modifies the object of study and its scientific understanding. For example, humans can and have changed their modes of social organization throughout history.

Thus, the understanding of social modes of production and the practical transformation of modes of production are mutually conditioned not only on roughly the same time scales but also in terms of the magnitude of change in the various forms of human social organization. All of this has changed dramatically with the Anthropocene crisis of habitability. For the first time in human history and in the history of the Earth, human action on the natural system can no longer be neglected, and now capitalist human action and the dynamics of the Earth are roughly coupled in terms of both the magnitude and the time scale of change of both the human system and the Earth system.

4.1.3 Natural Laws and Social Production Laws

There is, however, a more essential and not merely formal difference between the natural and social sciences. Natural systems do not evolve according to a teleological design that determines the ends to be achieved and the means necessary to achieve them. Rather, the evolution or history of natural systems is shaped by the interaction of multiple determinations that are mutually concatenated in a concrete way and, for this reason, they can be objectively known. Understanding the system means knowing the specific interplay of each determination in relation to the network of determinations that shape the object of study. That is, it means understanding why and how one determination plays a specific role and not another in a given system. For example, in the atomic system, each particle has a particular relationship to other particles, and only that relationship is possible within the limits of the system. Similarly, in the cellular system, in plate tectonics, each element also has a concrete relationship with other elements of the system. That is why deterministic laws describing the interactions of the elements forming the object and its history can be formulated. Therefore, the scientific-theoretical understanding of nature, if correct, reveals a number of deterministic laws and principles that are compulsory within the limits of each particular system. For example, Newton's law of gravity for macroscopic bodies, Heisenberg's principle for atomic particles, genetic laws for living beings, Steno's law of superposition for sedimentary strata, etc.

At each stage of the evolution of matter, there are new laws which are constitutive of that stage of evolution and which do not constitute the laws of previous stages, but conversely, the laws of previous stages are always constitutive of new stages of the evolution of matter. This means that the laws of inorganic matter are also the laws of living matter, including humans, but human laws of social production do not constitute the laws of inert matter. In fact, the constitutive laws of earlier and general stages of matter evolution take on new forms in the newer and particular systems of matter, i.e., such laws appear to be mediated by the specific structures of the new systems. For example, the laws of fluid dynamics of inorganic systems, such as viscous magma, take on new expressions in organic systems, such as viscous blood, which are mediated by the internal organization of living systems. Likewise, in the capitalist regime, the laws of genetics appear not only in the new form of genetically modified organisms, but also in the form of commodities produced for the exchange value inherent in every commodity, i.e., in the form of transgenic seeds mediated by the process of reproduction of capital. The phenomena of the material world, whether social or natural, are universally interconnected through multiple determinations. Understanding the laws and concrete mediations that govern these interconnections is the task of science. This can only be achieved from a monist epistemology and on a materialist and dialectical basis. In this way, the different forms of matter are seen as evolving from a common material ground, and the different laws and transitions between the stages of matter evolution can be understood (Rubinstein, 1963).

Therefore, humans are always subject to the laws of the natural systems regardless of the kind of the practical interaction with the system and regardless of the way in which the system is thought. Again, practice is always the criterion that validates whether a theory is correct or not. For example, nuclear energy and nuclear weapons validate particle physics as a correct theory and genetically modified organisms validate modern genetics as roughly correct. Note that such practical validation of theories does not imply an ethical position on the issues involved, nor does it have any immanent imperative. That is, the scientific understanding of particle physics does not necessarily mean that the atomic bomb must be built, nor does the scientific knowledge of genetic laws mean that living organisms must be genetically modified. Here, concrete ethical decisions are possible precisely because the scientific theories are roughly correct. In fact, ethical decisions permeate the entire process of scientific research, mainly in the form of ethics and morals that are currently considered standards for a given society. For this reason, scientific and technical knowledge is not neutral and always unfolds within a given historical determination that marks the structural trends of scientific development (Piedra Arencibia, 2018). In any case, the laws of nature are always obligatory and, for example, if humans want to fly with airplanes, they must strictly consider the law of gravity.

Moreover, in their practical interaction with nature, human beings must follow the laws of nature because they "can work only as Nature does, that is by changing the form of matter" (Marx, 1967a, p. 43). Social systems, however, can be changed by conscious and unconscious human action, and thus the specific rules and principles of one type of society can be replaced by others. This usually implies a profound transformation of a society into a new one, affecting the way in which social systems produce and reproduce themselves according to their particular features, and the way in which social reality is perceived and understood. For example, the form in which hunter-gatherer societies reproduced themselves, including both the material and ideal forms of social reproduction, is completely different from the form in which slave societies did so, which in turn is completely different from the particular form of feudal societies, and again different from that of capitalist society. All of them, however, share labor as a universal concrete activity that mediates between Nature and humans and without which no kind of social system is possible. Therefore, the key difference between natural and social systems, involving practical activity, epistemology, and ethics, is that social systems and their corresponding laws can be transcended whereas natural systems and natural laws cannot.

Capitalist production is planned at the scale of each firm, where costs and benefits are examined in great detail internally. At the scale of social reproduction, however, the capitalist mode is the typical case of an autonomous social system that develops spontaneously on the basis of commodified interpersonal relations that are not subject to any prior subjective design. Since the only teleological design of capitalist production is confined to the interior of individual firms based on profit, deterministic laws similar to those of natural systems can be discovered and describe the dynamics of the system. However, this requires an appropriate epistemological paradigm, otherwise the laws of capitalist production will not be revealed.

4.1.4 *Epistemological Paradigms of the Capitalist Society*

Capitalism is the first global society in human history, and it involves not only the material production of goods and services but also the ideal production, that is, the production of ideology, of notions, concepts, and theories. Material production and the production of ideas are necessary and inseparable moments for any social organization to reproduce itself according to its own foundations. The relationship between material and ideal production was clearly established by Marx and Engels on a materialist basis and allows us to distinguish an idealist conception of reality from a materialist one:

> "The phantoms formed in the human brain are also, necessarily, sublimates of their material life-process, which is empirically verifiable and bound to material premises. Morality, religion, metaphysics, all the rest of ideology and their corresponding forms of consciousness, thus no longer retain the semblance of independence. They have no history, no development; but men, developing their material production and their material intercourse, alter, along with this their real existence, their thinking and the products of their thinking. Life is not determined by consciousness, but consciousness by life". (Marx & Engels, 1976, p. 38)

There are two main epistemological approaches to understanding capitalism in particular and social modes of production in general. These approaches are antagonistic, and they are ultimately related to the cardinal problem of philosophy: "The great basic question of all philosophy, especially of more recent philosophy, is that concerning the relation of thinking and being" (Engels, 1946, p. 21).

4.1.4.1 *Positivism and Idealism*

The positivist view of human knowledge is closely related to liberal philosophy. In this view, knowledge is understood as an aggregation or sum of more or less random actions of individuals who carry out their activity as isolated monads with respect to the process of research. The social character of knowledge results from the aggregation of individual contributions once they are submitted to society, which validates the individual contributions a posteriori. In this view, individuals make inductive abstractions of generalities from empirical observation and sensory perception of phenomena, and then elaborate concepts, which are the verbal forms, the terms, describing the common features observed in the studied object.

Thus, positivism does not distinguish between notions—the forms of knowledge obtained through the inductive abstraction of generalities from phenomena—and concepts—the forms of knowledge that imply further abstraction and theoretical elaboration through deduction, or, if it does, the distinction is only formal.[7] Moreover, from a positivist perspective, there is no difference between thought and language, between the forms of logic and the laws of thought and their symbolic expression in language. Language is both the formal expression of concepts, judgments, and other forms of reasoning, and the main vehicle for the process of reasoning, but it is by no means thought as such.[8] The epistemological research of neo-positivism in the twentieth century on the procedure for validating scientific theories was based in part on the equation of thought and language. For example, Carnap's principle of verification states that a scientific theory is correct if it is in formal agreement with observed empirical data. In this way, scientometrics, which is based on the fundamental fallacy that the quality of science can be measured, and that it can be measured by the quantity of scientific publications, is validated as a correct theory. However, positivism does not examine the fallacy as such, but rather assumes it as a premise of the theory. Since, based on this assumption, the quality of science is empirically supported by all the indices and mathematical tools explicitly developed for the purpose of supporting the fallacy, the theory is in formal agreement with empirical data and is verified. The main logical principle underlying positivism is the principle of non-contradiction, which states that although reality appears to be contradictory, scientific theories cannot have internal contradictions and therefore must be formally consistent.

The dominant social theories in bourgeois society are based on positivism. In practice, this means that the studied object—in our case the capitalist mode of production, but in general any social mode of production—is considered a puzzle object composed of separate sections, e.g., education, economy, law, etc., which can be approached independently, and that the totality of the studied system results from the aggregation and interaction of the different parts. The *intersection* narrative developed in postmodernity is an example of this approach. Thus, the capitalist

---

[7] For a further development of the distinction between notions and concepts, see Ilyenkov (1982a).

[8] For a further development of the role of language in logic, see Rosental (1962).

system is not seen as a concrete totality shaped by multiple determinations that can be understood on the basis of fundamental concepts and categories that are intertwined in terms of causality and necessity and have been abstracted and unfolded from the reality observed in phenomena. That is, the positivist approach does not allow for the unfolding of fundamental principles or laws immanent in the system, whose historical development provides the ontological history of the system. Rather, the positivist approach proceeds through the inductive abstraction of generalities from the observed phenomena and from the interconnections between already established notions and analytical categories, which are usually assumed as premises without critical research in terms of the structure of the system. From there, a formal deduction is made that leads to the formulation of formally consistent theories. In this way, the history of the system is a phenomenological history, a historiographical account of events, notions, and categories, that is, of the phenomenal expressions of the social system and their relationships, which does not reveal their internal connections with the fundamental principles. Examples of social science disciplines driven by such a positivist understanding of the reality under study are economics, Malthusian and neo-Malthusian demography, and scientometrics.

Economics deals with the study of the material production of goods and services that all societies need to produce in order to reproduce themselves based on their particular characteristics. In the capitalist mode, production is not designed according to a previously conceived plan aimed at evaluating the needs of society and the means and processes necessary to satisfy these needs. Rather, production is first and foremost the production of commodities, and it is driven by the profit inherent in every commodity. On this basis, individual capitalists find their position in the productive process of society and receive a portion of the total profit produced. The spontaneous and unplanned interaction of individuals in the capitalist mode has been explicitly recognized by classical political economists such as Adam Smith with his famous invisible hand market, by economists of the Austrian school such as von Mises, Hayek, etc., who recognize the formation of prices as an aleatory process and the price system as a spontaneous mechanism providing information to individual capitalists for their investment strategies, and in fact, it is implicitly recognized by all economists regardless of the school to which they belong. Time series of stock market values have been mathematically modeled by polynomial fitting and other statistical techniques to provide investment decision tools for financial capitalists (Hendry, 2004). This is an example of the spontaneous and uncontrolled character of market prices, which are formed as a result of aggregated individual actions performed unconsciously with regard to a comprehensive understanding of the economic system. For this reason, not only a system of market prices is needed, but also mathematical tools aimed at navigating within this system. This example illustrates the positivist approach that underlies mainstream economics and, consequently, the misunderstanding of capitalist fundamentals. Polynomial fitting of continuous time series of gases released in fumaroles of active volcanoes is also used to infer processes occurring in the deep magma chamber that cannot be directly observed. However, while a volcanic system is entirely produced by nature, the capitalist system of market prices is entirely produced by the human hand, and the need for similar mathematical approaches to make inferences about both systems clearly shows that the system of capitalist production is beyond human control.

The positivist perspective conceives the capitalist mode ahistorically, as an absolute mode of production, rather than as an organic system resulting from the internal development of fundamental elements that constitute the essence of the system and whose contradictory unfolding reveals the empirical phenomena observed and the history of the system. Positivism does not investigate the connections between the phenomenal forms (prices, profit, rent) and the essential concepts underlying these forms (commodity, labor, value). As a result, the system is not studied as an organic whole, but only the observed phenomena are considered in order to articulate a formally consistent theory that ends up being a tautology, usually based on argumentative fallacies. In this way, the system under study is not subjected to critical research, but, on the contrary, it is formally legitimized. For example, the Austrian school supports the need for a free market and the formation of aleatory prices by assuming that production can only be organized on the basis of aggregated individual actions driven by profit. That is, a historical form of social production and its corresponding social being is taken as absolute and, according to this premise, the characteristic economic interaction of individuals of such a historical form is given as an argument for the institutional expression of this interaction, namely the system of market prices (Cockshott & Nieto, 2017).

The scientific understanding of the capitalist social system by mainstream economics is reduced to a formal explanation that absolutizes the capitalist social being and the phenomenal forms of the capitalist mode of production. This has a straightforward epistemological corollary: the materialism underlying this understanding of the capitalist mode is a positivist-based materialism, and therefore it leads to an idealist-based understanding aimed at legitimating this particular mode of production. Since the fundamental contradictions of this particular social system, as shown by the recurrent economic crises, the unequal distribution of wealth, and the Anthropocene crisis of habitability, etc., remain unexplained and unresolved, the understanding of capitalist society by bourgeois economics cannot be considered scientific. Most mainstream economic schools—marginalist, neoclassical, Keynesian, neokeynesian—share the common background of a subjective theory of value, which implies that value is an individual perception that cannot be known objectively, and that only a formal, mathematical understanding of value forms (rent, profit, capital, etc.) is possible. The differences between the various mainstream schools of economics are therefore only formal differences in the subjective perception of value. As a result, an empirical phenomenon of capitalist production, such as the periodic economic crises, is always attributed to the aggregation of bad individual decisions, but the immanent relationship of such crises to the essentials of the mode of production is never investigated.

4.1.4.2 Dialectics and Materialism

An opposed epistemic view of positivism is that of dialectical materialism. Here, knowledge is understood as a social and historical process concerning the whole humanity, which is confronted by individuals at any given historical time and to which individuals contribute with their living activity. In this view, the historical process of understanding reality is not seen as a sum of the contributions of individuals in every epoch, but as a collective process in which previous knowledge is critically assimilated within a new one and is understood on a materialist basis. That is, in terms of the material conditions that made former knowledge necessary at a given historical time and that allow to understand it in the current time and within the whole historical evolution. This is the way Marxists weigh the contributions of classical political economists to the concept of value and to the labor theory of value. Similarly, although Marx, Engels, and classical Marxists are usually classified as atheistic, the fact is that they understand religion as the necessary form of fetishism for the period of human knowledge in which reality could not be understood on its own basis yet.⁹ In this regard, God and Gods are completely real and necessary products of the human mind in order to understand reality at a given historical time.

On a dialectical and materialist basis, notions are seen as social rather than individual, and they are fixed and expressed in language through historical and social processes, not as a result of individual perceptions and abstractions. Rather, individuals come into historical social systems with already fixed notions and confront them as an objective reality to be individually grasped and appropriated. In this view, thought is seen as the particular mode of ideal activity of the social being as reflected in individuals. From a dialectical and materialist point of view, notions are fixed in culture and they are distinct from concepts, which result from the necessary theoretical elaboration of notions through abstraction and logical deduction in thought and have a symbolic expression in verbal language, mathematics, music, etc.

Marx's Capital is the best example of a materialist and dialectical approach consciously and systematically applied to the understanding of the capitalist mode of production. In other words, Marx's Capital is a practical exercise of a dialectical and materialist epistemological view aimed at understanding a concrete and historical mode of social production. However, after the collapse of Soviet-style economies, dialectical materialism has not been well received in Western culture, even though this epistemic view reached its highest scientific development in the Soviet Union after the Second World War. In recent years there has been renewed interest in Soviet philosophers of this period, especially Evald Vasilievich Ilyenkov, but most Soviet philosophers and psychologists of the second half of the twentieth century are still little known, and many of them have not been translated into Western languages. As a result, there is a remarkable epistemological gap between Western and Eastern cultures that hinders a fluid dialogue and is ultimately based on two contrasting epistemic views, a positivist and idealist understanding of social systems in the West, and a materialist dialectical one in the former Soviet East (Chukhrov, 2013).

In Marx's Capital, capitalist society is examined as an organic historical system through the unfolding of the internal contradiction of the fundamental element of the system, the commodity. By developing the

---

⁹ In essence, this is what Marx meant by criticizing Feuerbach in the Fourth Thesis on Feuerbach.

contradiction inherent in commodities, Marx is able to understand value as a historical form of the products of labor, money as a phenomenal form of value, capital as the form of value that valorizes with the exploitation of the surplus value exerted by labor, and capital as the fundamental social relation of capitalist society. The phenomenal forms of value (salary, rent, profit, interest, etc.) are logically deduced in Marx's Capital through the sequential unfolding of the contradictions of the analytical categories derived from the practice of capitalist production, starting with the commodity as the primary element. Here, each abstracted concrete concept accounts for the necessary conditions of existence of each particular phenomenal form empirically perceived in relation to the rest of the phenomena under study. In turn, each abstracted concrete concept is the necessary consequence of the existence of the rest of the phenomena. Thus, in Marx's dialectic and materialist understanding of the capitalist mode, profit is a form of value, it is both a consequence of the development of the internal contradictions of the commodity form of value and a prerequisite for the valorization of value and for other forms of value such as rent. The result of Marx's research is a theoretical corpus in which abstract and simple categories are dialectically connected to complex and concrete categories, allowing us to understand not only the material production but also the ideal production, that is, how the ideal forms are settled in the culture and language of bourgeois society. Such a concrete interrelation of the multiple determinations that shape the particular object of study cannot be achieved through the inductive abstraction of generalities from phenomena. It can only be achieved through a deductive-logical process in which inductive abstraction is a necessary vanishing moment. This means that the concept of value cannot be obtained by abstracting generalities from the forms of value, but only by unfolding the contradictions of the fundamental element of the system, the commodity, through logical deduction.

Political economy is the science that studies the material production that any society must undertake in order to reproduce itself, and that recognizes a political character to the organization of such production. It is quite different from today's economic policy, which deals with the management of social production in a mode of production that is taken as absolute, such as the capitalist mode, and therefore lacks a critical apparatus. In his critique of classical political economy, Marx followed an epistemological approach to the capitalist society that benefited from modern empiricism and materialism—Bacon, Locke, Diderot, Helvetius, etc.—which he critically reviewed by examining Feuerbach's materialism. Marx also benefited from the modern dialectical tradition, beginning with Spinoza and culminating with Hegel, which he also critically reviewed.[10] In this way, he was able to go beyond the classical political economists of modernity—Petty, Smith, Ricardo—in the scientific understanding of the capitalist mode. The classical political economy reached its highest development with Ricardo and started to be demolished by James Mill, who conducted an understanding of the economy in terms of formal logic, banning contradictions in the theory of value and requiring formal concordance among the concrete determinations of the theory (Ilyenkov, 1977). However, Ricardo failed in his attempt to build a theoretical corpus aimed at organically understanding the fundamentals of capitalist production. First, he conceived capitalist production as absolute and not as a historically determined mode of production. Second, he failed to identify the primary element, the commodity, from which the other phenomenal forms of the mode of production unfold. Although Ricardo correctly distinguished concepts as concrete theoretical abstractions from the notions fixed in language, he imagined an immediate correspondence between value and the forms of value, that is, a non-contradictory correspondence between value and profit, for example. As a result, he pretended to find value by abstracting it from the generalities of the forms of value (profit, salary, rent) and, as long as this was not possible, he concluded that the unfolding of value into forms of value occurs only as an abstract deduction in thought, and not in reality. In summary, Ricardo's efforts to derive value from value forms were based only on formal deduction, revealing the positivist and idealist nature of his approach (Ilyenkov, 1982a).

Nevertheless, Ricardo's efforts to understand the relations between the universal concrete concept of value and its phenomenal expressions were much more scientific than those of most of the bourgeois economists who followed him. At least he correctly identified abstract labor as the substance of value and production as the key moment of the valorization of capital, whereas most current mainstream economists engage in a scholastic legitimation of the capitalist mode, usually based on vulgar empiricism. Piketty's bestseller, for example, presents impressive data on

---

[10] "Spinoza ... showed Man's nature as a peculiar form of nature in general, and their dialectic understanding both as the unity and as the difference within this unity ..." Directly translated into English from Iliénkov and Naúmienko (1977).

the unequal distribution of wealth in capitalism, but he is unable to link it to the structural foundations of capitalist production (Piketty, 2014). As a result, his proposal is reduced to simply taxing the profits of capital in order to redistribute wealth on the basis of such taxation. Such a simple proposal obviously does not require a real scientific understanding of the mode of production and does not examine the structural contradictions of capitalist production. Moreover, the mathematics developed by Piketty is based on neoclassical fallacies aimed at showing that the rate of profit of capital does not decrease during capitalist production and, therefore, that the reproduction of capital is capable of producing infinite wealth, which, unfortunately, is wrongly distributed. Marxist research using the same database as Piketty has shown that the rate of profit does fall when it is properly estimated (Maito, 2014). Based on an epistemic paradigm opposite to Piketty's, Marx was able to show that unequal distribution of wealth is a must of capitalist production that no tax can cancel. Marx did not just describe empirical data like Piketty, but based on logical deduction following the law of value, he formulated the law of capitalist accumulation, which has this succinct expression: "Accumulation of wealth at one pole is, therefore, at the same time accumulation of misery, agony of toil slavery, ignorance, brutality, mental degradation, at the opposite pole, that is, on the side of the class that produces its own product in the form of capital" (Marx, 1967a, 1967b, p. 345).

4.2 The Anthropocene and Capitalocene as Concepts of the Planetary Crisis

The realization that we are living in a critical historical moment regarding the conditions of habitability on Earth, not only for humans but also for many other living organisms, is gaining ground among ordinary people, scientists, politicians, and social movements. This critical time has been characterized as the planetary crisis of the Anthropocene, and studies conducted in this century show that habitability on Earth is progressively deteriorating. These studies underscore the need for a profound rethinking of the current relationship between humans and the rest of the planet. Such a radical rethinking of our role on the planet must necessarily involve the whole of global society (Palsson et al., 2013). Accordingly, most researchers from the natural and social sciences call for the integration of the social sciences and humanities and the natural sciences into Earth system science, with the aim of building a broader and more comprehensive theoretical corpus capable of addressing the habitability crisis of the Anthropocene.

Until now, however, there has been a dual approach to the habitability crisis on Earth, on the one hand from the social sciences and humanities, and the other hand from the natural sciences. Such a dualism has led to some disagreements and disputes between the social and natural science disciplines regarding the formalization of the habitability crisis in the Geologic Time Scale, but not only (Ellis et al., 2016; Thomas, 2024). Many scientists from natural science disciplines regret the difficulties in dealing with social science theories, due to the lack of overall consensus on the topics studied and the numerous and dispersed approaches in the social sciences, which hinder fluid communication and proper understanding between natural and social scientists (Mooney et al., 2013; Oldfield, 2018).

There is also a growing, albeit more limited, perception of the close relationship between the ongoing habitability crisis and today's global capitalist society. This perception is based more on intuition and the historical correspondence of indicators of the habitability crisis with the capitalist mode of social production than on scientific studies showing that the crisis is a structural necessity of capital reproduction. As a result, a number of alternative terms to the Anthropocene have been coined to name the current historical time. While terms such as *Plantationocene*, *Chthulucene*, *Growthocene*, *Econocene*, *Pyrocene*, *Necrocene*, and other similar terms may have a provocative scope, it is also true that they are based on an incomplete understanding of the ongoing crisis. Among the alternatives to the Anthropocene, the Capitalocene is the term that has undergone a deeper conceptual development. However, both the Anthropocene and the Capitalocene concepts are not free of important misconceptions about the crisis and its relation to the foundations of the capitalist mode of social production based on the reproduction of capital.

Emerging from the planetary crisis requires a scientific understanding of the functioning of the Earth as an integral natural system, and this requires the involvement of many disciplines of the natural sciences. It also requires a scientific understanding of the different modes of social production throughout history—especially the capitalist mode—and their specific impacts on the functioning of the planet. Today, the plethora of terms and the different conceptualizations that underlie them, which usually reflect only partial aspects of the planetary crisis, have created some confusion that hinders both a proper understanding of humanity's role in the crisis and our ability to adopt the right strategies to get out of it. Since concepts are verbally expressed through terms—although not all terms are concepts—it becomes clear that the current Anthropocene-Capitalocene debate is not just a terminological issue. It is a debate about the conceptual content of the crisis underlying these terms and about the different approaches to addressing it, and this has important implications for the future of humanity on Earth, as well as for the future of the Earth, because misunderstandings of both the planetary crisis and the fundamentals of the capitalist mode of production, and of their intertwined nature, are ultimately responsible for the weakness of the policy proposals for overcoming the crisis.

To some extent, the Anthropocene-Capitalocene dispute illustrates the disagreements between the natural and the social sciences, which are usually related to different views on the concept of science and scientific knowledge, and on the possibility of formulating deterministic laws for the social realm similar to the laws of nature. The Anthropocene is a concept closely linked to the development of modern geosciences and Earth system science, while the Capitalocene and other *-cene* terms are concepts developed within the humanities and social sciences. Their disagreements about the habitability crisis are related to different conceptualizations of the crisis and ultimately have an epistemological background. For this reason, it is necessary to examine the concepts of the Anthropocene and the Capitalocene from an epistemological perspective. The focus is mainly on the dispute between Anthropocene and Capitalocene, both because these are the more widely used terms and because they have undergone deep theoretical development.

4.2.1 *The Epistemological Perspective of Earth System Science on Nature and the Habitability Crisis*

The Anthropocene is a concrete concept, a concrete abstraction shaped by the synthesis of multiple determinations interacting from both the natural and social realms. The concept of the Anthropocene as a planetary crisis is obtained by theoretical abstraction and logical deduction from sensory perceived phenomena and empirical experience. The empirical data collected by Earth system scientists on the Anthropocene crisis are extremely clear: the current planetary crisis is human-induced, it is historically limited to the last 200–300 years at most, and the crisis itself, as shown by the ensemble of empirical indicators, has deepened during this historical period, regardless of whether a particular indicator shows a partial recovery. When the socioeconomic roots of the habitability crisis are examined on a dialectical and materialist basis, the Anthropocene crisis reveals that human action, exercised through a particular and historically determined social system, that is, the capitalist mode of social production, is the main determinant in the concrete conception of this planetary crisis.

The habitability crisis of the Anthropocene has altered the traditional neo-Kantian divide between the natural and social sciences. The Anthropocene, as a concept arising from human interaction with the planet under a historical mode of social organization, reveals above all the inadequacy of the traditional dualist approach based on the separation of the natural and social sciences. Everything is now so closely interrelated that such a dualist approach is methodologically flawed. Moreover, the close interaction between natural and social processes cannot be fully understood if natural and social spheres are treated separately. Second, the habitability crisis shows that scientists can no longer approach the Earth system as external observers, as has been the case in the natural sciences. Finally, it shows that practical action to overcome the planetary crisis depends on a correct understanding of the social and natural determinants of the crisis. That is, if the interrelationships between natural and social determinations are not properly conceptualized under an appropriate epistemic paradigm, the crisis will deepen and humanity will not be able to transcend it.

Natural and social science disciplines differ formally in the time scales over which the objects studied and the theories about them evolve. In addition, natural systems and natural laws are not substitutable, while social systems and their laws can be changed by others, provided an appropriate human action is taken. For these reasons, and because human beings are self-produced social beings, philosophers and thinkers after the neo-Kantian split between the natural and social sciences have questioned whether it is possible to undertake a scientific and objective understanding of social systems, as is usually undertaken for natural systems. When Earth system scientists approach social science theories in the context of the Anthropocene crisis, they are, perhaps inadvertently, confronting this epistemological question. The epistemological perspective that Earth system scientists adopt in addressing this question is crucial to a successful understanding of the Anthropocene dilemma.

Understanding nature implicitly acknowledges that nature is an objective material reality outside the human mind and independent of the way it is thought and known. Earth system science is no exception to this recognition. Understanding the Earth as a concrete entity, in which the particular interactions of multiple determinants are to be characterized, is always an unfinished process that allows us to improve our knowledge of the history and dynamics of the Earth system. Our current understanding of the Earth would not have been possible from an idealistic perspective, where the Earth lacks any concrete internal articulation and only individuals are capable of giving structure to nature through thought and language. In other words, from such an idealist perspective, nature is deprived of any causal concatenation or history, which is seen as merely an attribute of human thought and language (Piedra Arencibia, 2019). From a materialist perspective, a materialist approach to nature is a requirement of the object studied. It is an epistemological requirement based on the fact that nature itself is an objective and material reality. Earth system scientists, whether they are aware of it or not, are proceeding from a materialist epistemological perspective when dealing with the natural component of the Anthropocene crisis of habitability.

Nature is a material and dialectical entity, and the dialectics of nature is not a theoretical construct or metaphysics, rather the dialectical understanding of nature by thought is the necessary reflection of the dialectics inherent to nature.[11] Otherwise, how is it possible to construct dialectics if it is not yet present in the object studied, which is the ultimate source of any empirical perception? Such a construct would be necessarily wrong if it could not reflect the true constitution of reality. The origin of life and the evolution of species are examples of dialectics in nature. Organic molecules, which are the basis for the origin of life on Earth, evolve as a negation or contradiction of inorganic molecules and constitute a new structure of matter with new and special laws that are different from the laws of the inorganic realm. Contradiction is also the driving mechanism of the evolution of species, in which differences between individuals within a species evolve into differences between species. That is, the contradiction of some individuals with respect to the species average differentiates as a new average that negates the former

---

[11] This is essentially Engels' epistemological position (Engels, 1986). However, Engels' contribution to the so-called theory of reflection and dialectics has not only been misunderstood, but even ridiculed in Western Marxism (Piedra Arencibia, 2019).

one.[12] The internal constitution of matter, from the Higgs boson to the classical atomic particles and molecules, is the result of an evolution based on the contradictions of the constitutive elements, usually expressed in terms of opposing electric charges, but not only. In fact, particle physics is based on the principle of contradiction as the driving mechanism by which matter is constituted and differentiated. The well-known wave-corpuscle duality is an example of unity based on a contradiction that defines both the identity and the difference of matter. The practical application of the wave-corpuscle duality in electron microscopes shows that the contradiction is real, and for this reason, it is shown as a logical contradiction in thought and is expressed in language. The elliptical orbit of planets and satellites is defined as the unit of two opposite motions by which planets approach and move away from a larger planet or star. Thus, the dialectical understanding of nature reflects the historicity of nature, the evolution of matter from inorganic matter to organic matter, to living organisms, and humanity as the "thinking body of nature" through which nature is able to think about the process of thinking, and as claimed by Spinoza thought becomes an attribute of nature (Engels, 1986). Such a dialectical understanding reflects the autonomous evolution of nature by overcoming successive contradictions, in which at each stage of evolution the newly established conditions assume at their core the structural laws and evolution of previous stages. For this reason, if the understanding of nature is roughly correct, the contradictions of nature must be reflected in the forms of thought, as contradictions of logic.[13] To borrow a term from evolutionary biology, the evolution of natural systems occurs in homeostasis, and thus inorganic matter evolves within the structure of organic matter, cellular elements evolve within the structure of more complex living organisms, and so on.[14] As shown by the Anthropocene crisis, the entire Earth system is now evolving within the structure and laws of the capitalist mode of production. Hence, this crisis reveals the ultimate

---

[12] Evolutionary biologists such as John Haldane, Stephen Gould, Richard Levins, Richard Lewontin, John Bernal, Faustino Cordón, and others, have correctly grasped the dialectics in nature (Foster, 2022).

[13] Rosental (1962) makes an extended analysis on the relationship of contradictions as categories of logic and of the dialectical reality, and on the different treatment of contradictions by formal logic and dialectical logic in the formulation of concepts and judgments.

[14] For the concept of homeostasis in evolutionary biology, see Cordón (1982).

contradiction of the capitalist mode: the contradiction of the reproduction of capital with the reproduction of nature and with humans as the thinking body of nature.

However, most Earth system scientists are unaware of the dialectical constitution of nature, just as they are unaware of the dialectical and materialistic epistemological perspective they follow in their scientific understanding of nature. Scientists in the natural sciences are forced to spontaneously follow this epistemological perspective because the object of study is a material and dialectical reality. The epistemological approach to nature must be consistent with the structure of nature, otherwise knowledge is incomplete. In other words, the dialectic of nature is reflected in human thought as dialectical logic and as dialectical and materialistic understanding. Yet, as has been pointed out paraphrasing Marx, naturalists "do not know what they do but they do it" (Piedra Arencibia, 2019). This is the key difference between Marx's approach to the study of social modes of production and naturalists' approach to the study of nature. Marx consciously adopted a materialist and dialectical view to understand capitalism, which he systematically applied in the critique of political economy as shown in Capital, while naturalists have unconsciously adopted this epistemological view to understand nature. Nevertheless, the conscious Marxian dialectic and the unconscious naturalist dialectic are the correct approaches to understanding the objects studied, as shown by the predictive character achieved by their respective scientific corpus. For example, the astronomical calibration of deep-sea sediments not only allows the time scale of the chronostratigraphic successions of the geological record to be calibrated but also allows the cyclic stratigraphy of future successions to be predicted, given certain boundary conditions. Similarly, the law of the tendency of the rate of profit to fall predicts the long-term evolution of capitalist production and has been empirically confirmed more than a century after Marx's formulation of the law.

4.2.1.1 Limitations of the Earth System Science Understanding of the Habitability Crisis

Earth system science inadvertently adopts a materialistic and dialectical perspective in understanding the present configuration of the Earth and its history. Based on this approach, a more concrete and complete knowledge of the multiple determinants and their interrelationships that shape the Earth system is now possible. Consider the evolution of the

Bretherton diagram or the Geologic Time Scale from their initial configurations in the last century up to the present day. Collectively, they show the ongoing process of a more complete and concrete understanding of the Earth. Unfortunately, this applies only to the natural side of the object of study, not to the social or human side. With regard to the social aspect of the Earth system and the ongoing crisis of the Anthropocene, Earth system scientists operate from a positivist and idealist point of view, and as a result, a scientific integration of humans and their social modes of production into Earth system science becomes impossible. Earth system science promotes stewardship of the Earth to address the habitability crisis, but it is based on a positivist and idealist view of the relationship between humans and nature, and on a misunderstanding of the fundamentals of capitalist production. Ignorance of the socioeconomic laws of the reproduction of capital leads to an idealist conception that imagines that humans control the mode of production, when in fact the opposite is true: they are controlled by a mode of production in which capital is the subject that dominates social reproduction. Therefore, the human agency imagined by Earth System scientists to face the crisis of the Anthropocene is a fallacy within the limits of capital reproduction, because the only possibility for a harmonious relationship between humans and nature lies outside this mode of production.

Earth System science is claimed to be a "transdisciplinary endeavour aimed at understanding the structure and functioning of the Earth as a complex, adaptive system," and has "the grand challenge…to achieve a deep integration of biophysical processes and human dynamics to build a truly unified understanding of the Earth System" (Steffen et al., 2020, p. 54). However, Earth system science is not only far from meeting this "grand challenge," but it is moving in the wrong epistemological direction. As a result, the conception of the planetary crisis by Earth system science is incomplete, and proposals to face the crisis are insufficient, to say the least. The main reason is because the studies conducted by Earth system scientists have failed to identify the concrete anthropogenetic causes driving the planetary crisis, thus, the crisis is assigned to humans in general. In this way, Earth system science erases any history regarding the forms of social production and reproduction, and does not consider that there are different historical modes of social production, with different social metabolisms and different impacts on Earth. All of the environmental indicators of the planetary crisis reveal a clear departure from their historical background level within the last two centuries.

This marks an undeniable historical correlation with the capitalist mode of production and not with humans in general. Although the historical coincidence of the planetary crisis and the capitalist mode is not evidence by itself of a structural or immanent link between the two in terms of essence, necessity, and causality, it indicates the correct direction to be followed for any research aimed at building up "a unified understanding of the Earth System," including its habitability crisis, and at leaving this crisis behind. In other words, if the crisis is historically correlated to the capitalist mode of social organization, any research that claims to be scientific should investigate the possible link between this mode of production and the ongoing planetary emergency. However, it is a paradox that Earth system science disregards the empirical evidence built upon its own studies, which clearly correlates the planetary crisis to the capitalist mode of production and not to previous modes, and that the possibility of such a structural link in Earth system science studies is not even mentioned. The question is: why?

There is not a simple answer to this question. Earth system scientists are certainly aware of the unequal contribution to the habitability crisis of the various countries and regions of the world depending on the development of capitalism, and also of the unequal contribution of different social classes. Therefore, Earth System scientists may suspect that the crisis is somehow related to the capitalist mode of production, in a similar way that workers suspect that someone else is being enriched with their work. However, empirical data on the differential contribution of classes and countries to the crisis do not explain why and how the crisis is a result of capitalist production, and why and how the differential contribution is ultimately related to the commodification and division of labor at a local and global scale. Farther scientific research aimed at unfolding the objective processes by which the habitability crisis is necessarily linked to capitalist production, and at understanding why there must be such a differential class contribution on this basis, is needed to turn these intuitions into scientific evidence. Earth System scientists have not undertaken such research. There might be a rather prosaic reason for this: research needs funding. It is possible that Earth System scientists are not confident in their ability to raise funds for investigating the relationship of the habitability crisis with the capitalist mode, and that they have decided to remain in their comfortable field of knowledge without exploring this possibility. On the contrary, Marx was able to demonstrate that the planetary crisis is inevitable under the capitalist production by revealing the causal concatenations of the metabolic rift, as a precursor of the potential crisis of habitability, with the particular form of labor exploitation under the capitalist mode in the context of his labor theory of value.

Earth system scientists are generally unaware of the scientific character of Marx's theory of value. Studying the fundamentals of the capitalist mode in order to investigate whether the planetary crisis is inherent or not to this mode of production is certainly outside of the realm of the natural sciences, and often of the expertise of Earth system scientists. In fact, many scientists from the natural and social disciplines believe that social production in general, and capitalist production in particular, do not proceed under deterministic—if nonetheless historical—laws that are similar to the laws of nature. Marx's theory of value demonstrates the opposite. Earth system scientists claim that an "integration of biophysical processes and human dynamics" is needed for a scientific understanding of the Earth system and hence, of the planetary crisis. In practice, however, they reject incorporating the only truly scientific corpus available of the capitalist system, which is Marx's theory of value. By unfolding the essential contradiction of the value system Marx is able to show the finite nature of the capitalist mode. He concludes that under this production mode, human social metabolism is not only mediated by the reproduction of capital, but also the reverse: the social metabolism becomes alienated to humans by the reproduction of capital, insofar as human's metabolism becomes a mere means for capital's metabolism.

The planetary crisis expresses the capitalist fundamental contradiction in the form of a final dilemma: the capitalist system or the human system. However, the finitude of the capitalist mode appears to be somewhat inconceivable for Earth system scientists, and they seem to adhere only to the first part of the famous Fredric Jameson quote, "Someone once said that it is easier to imagine the end of the world than to imagine the end of capitalism. We can now revise that and witness the attempt to imagine capitalism by way of imagining the end of the world" (Jameson, 2003, p. 76). As a result, the current understanding of the Earth's "functioning" by Earth system science incorporates the Earth's "biophysical processes," but excludes those views of the "human dynamics" that seriously question the validity of the capitalist mode for humanity and for the Earth's habitability. This is outside Earth system science's agenda. Claims that "collective human action is required" to stabilize Earth in habitable conditions and that "such action entails stewardship of the entire Earth System—biosphere, climate, and societies—and could include decarbonization of the global economy, enhancement of biosphere carbon sinks, behavioral changes and technological innovations, new governance arrangements and transformed social values" are not much more than a collection of vague statements and wishful thinking (Steffen et al., 2018, p. 8252). It is clear from this and other statements, and from the affinity between Earth system scientists and the field of ecological economics, which is strongly influenced by neoclassical economics, that the so-called global economy is nothing more than the capitalist economy, and that any other socioeconomic order than capitalism is automatically discarded (Costanza et al., 2012). While some stewardship of the Earth based on science will likely be necessary to mitigate the planetary crisis, it is easy to imagine that a stewardship determined by the reproduction of capital, in which profit is the only goal, could devolve into a kind of green-technological fascism.

Due to the immanent relation of the Anthropocene crisis to capitalism, a scientific understanding of the capitalist mode of production is unavoidable in order to properly address the planetary crisis underlying the concept of the Anthropocene. Otherwise, the practical actions undertaken will regrettably only alleviate some of the effects of the planetary emergency, but they will not be able to transcend it. When naturalists incorporate social science theories into the corpus of Earth system science they do not critically reexamine these theories, including those of the capitalist mode of production. Rather, they form their ideal conceptions of social issues according to the current and dominant epistemic paradigms in bourgeois society. Thus, with regard to the social component that needs to be considered along with the natural component in order to have a comprehensive understanding of the habitability crisis, the epistemological view adopted in Earth system science is that of mainstream economics, philosophy, politics, and so on. However, a positive solution to the crisis demands a critical reexamination of theories in most social science disciplines, and in particular of theories regarding the capitalist mode of production, because the ultimate cause of the crisis is the concrete relation between humans and nature under this production mode. Neither scientists from the natural sciences nor scientists from the mainstream social disciplines involved in Earth system science have undertaken such a critical review, and so they are not equipped with the appropriate epistemological background required by the Anthropocene crisis.

Scientific research on any topic develops under certain determining factors that correspond to a given historical society, including ideology, moral values, and technological means. Although these factors are themselves historical, they mark the structural trends and boundary conditions of research and only after long and arduous periods can they be overcome. Studies on the habitability crisis have mainly developed after the demise of Soviet-type societies and during a time of pompous proclamations about *the end of history* or that *There Is No Alternative* to the capitalist mode.[15] These narratives are firmly established in Western societies, to which most Earth system scientists belong. Following dominant Western thought, Earth system scientists have assumed the failure of these socialist experiences as an empirical demonstration of the impossibility of any mode of social organization other than the capitalist one. After all, in class societies, "the ideas of the ruling class are in every epoch the ruling ideas, i.e. the class which is the ruling material force of society, is at the same time its ruling intellectual force" (Marx and Engels, 1976, p. 67). Thus, Earth system scientists, whether they like it or not and whether they know it or not, are structurally conditioned in their research by the dominant bourgeois thought. This structural constrain is particularly evident when Earth system scientists deal with the social side of the problem of the habitability crisis.

In fact, not only Earth system scientists, but most thinkers and scientists in the West including those of the social sciences, have equated the failure of socialist experiences to the general failure of the only scientifically based alternative—which, paradoxically, bourgeois society itself historically gave rise to—aimed at overcoming the insurmountable contradictions of the capitalist mode of production. Without even performing a serious *concrete analysis of the concrete situation*, which would have revealed the internal political decisions that, among other things, led to the restoration of capitalism in the former Soviet Union, Earth System scientists have uncritically followed the dominant thought assuming the impossibility of any alternative mode of social organization to capitalism, as socialism and communism.[16] In this way, despite the undeniable advances in our current understanding of Earth dynamics thanks to the

---

[15] See Fukuyama (1992). *There Is No Alternative* (TINA) was a slogan coined by United Kingdom's prime minister Margaret Thatcher after the demise of the Soviet Union.

[16] See Lukács (1970).

development of Earth system science during recent years, this discipline has failed to account for the most crucial element in the ongoing planetary crisis. Namely, humans and their history, and in particular the most recent form of human social organization based on the reproduction of capital through commodity production. That is why Earth system scientists have a non-historical view on human history and the historical modes of social organization, and consider the capitalist mode as the absolute mode with which society can be organized and with which social reproduction is guaranteed.

Although it may sound paradoxical, the Anthropocene crisis as revealed by Earth system science studies is the empirical evidence that humans are following the correct epistemological path with regard to the understanding of the concrete functioning of nature. The Anthropocene shows that the materialist and dialectical view of nature is roughly correct because it reveals the true internal articulation and history of nature in the form of a planetary crisis. Like most natural sciences, Earth system science takes a dialectical and materialist epistemological view in understanding the natural side of the Anthropocene crisis. From the social side of the problem, however, the epistemological view adopted turns into a positivist and idealist one, which is dominant in the mainstream social sciences. In this way, the historical and concrete character of the Anthropocene is viewed from two opposing epistemological perspectives. On the basis of this dual epistemological approach, a comprehensive understanding of the Anthropocene dilemma within Earth system science becomes impossible.

From an epistemological and methodological perspective, this is a non-scientific approach, and accordingly, the resulting conceptualization of the planetary crisis by Earth system science is incomplete and not fully scientific. This critique of Earth system science is somewhat similar to Lenin's critique of Ernst Mach and Henri Poincaré, and Rosental's critique of Werner Heisenberg, Max Born, and other physicists for their idealism and positivism when facing problems regarding the epistemology of science and the relationship between thought and nature (Lenin, 1974; Rosental, 1962). As noted by Ilyenkov: "Not every artist has a well-developed concept of art, by any means, although he may create magnificent works of art. The present author is not ashamed to admit that he has a rather vague notion of the atom, as compared to a physicist. But it is not every physicist that has a concept of the concept" (Ilyenkov, 1982a, see Concrete Unity as Unity of Opposites). Earth system scientists certainly do not have a proper concept of what social reproduction is, what the laws of social production are in general, and what the laws of capitalist production are in particular. However, in defense of Earth system science, it can be said that not all Marxists have a well-developed concept of nature, because nature has been largely considered as non-dialectical in most Western Marxism, including the so-called philosophy of praxis and the Frankfurt school (Foster, 2013, Piedra Arencibia, 2019).

4.2.1.2 Complaints About Marxism from Earth System Science
Among the positive features brought to us by the Anthropocene issue is the open and radical character of the debates it generates. Frank Oldfield, founding editor of the journal *Anthropocene Review*, provides an example of such character in his review of book reviews (Oldfield, 2018). The essential points raised by Oldfield are that Marxism cannot address the social dimension of the Anthropocene challenge and that it precludes the necessary synergy between Earth system science and social science that is demanded nowadays. Oldfield reaches these conclusions based on book reviews on the Anthropocene topic—three books ascribed by him to Marxism—and on his own convictions. In doing so, however, Oldfield falls into some misconceptions about Marx and Marxism, which are understandable given that he has only an approximate idea about these issues, and which deserve some critical comments aimed at supporting his call for synergy between Earth system science and social science.[17] Although Oldfield's view of Marxism is a personal view, his view can quite likely be extended to that of many Earth system scientists, so the comments on Oldfield's conception of Marxism can also be extended to most of Earth system science.

According to Frank Oldfield, social scientists need to place their discourses on particular conceptual frameworks that seemingly exclude other approaches. Oldfield appears to assume that Earth system science or natural sciences in general do not adhere to particular conceptual frameworks, a claim that is debatable. For example, when analyzing a rock outcrop, structural geologists and paleontologists provide different interpretations because they observe different features on a given outcrop. Paul L. Hancock, founding editor of the *Journal of Structural Geology*, used to say that "you see what you know to see." Theoretical frameworks are not only necessary but are unavoidable for any understanding

---

[17] This critique does not pretend to be the particular defense of the authors' books reviewed by Oldfield, whose characterization as Marxists is somewhat doubtful.

of reality is pursued, whether it is natural or social. For geologists to understand the essentials of palaeomagnetism, for instance, it is necessary to approach an understanding of Newtonian and Quantum physics. Similarly, historians or economists should look to the Marxist tradition to understand the essentials of capitalism on a materialist and dialectical basis. On the other hand, Oldfield is right, and a Marxist approach to, say, economics excludes a bourgeois approach, and vice versa. But this is also true of the natural sciences, and in this respect, Oldfield lacks a historical perspective on the development of the natural sciences, for the history of the natural sciences is full of disputes, disagreements, and mutually exclusive paradigms. The beginnings of modern physics and astronomy were fiercely opposed by the dominant theories of the time, and as long as new theories implied a radical epistemological shift from the dominant worldview based on religious idealism, they were also fiercely repressed. For example, during the so-called Copernican revolution, which implied the change from a geocentric to a heliocentric theory. The repression by the Inquisition of Galileo Galilei and Giordano Bruno, among others, is well known. Toward the end of the nineteenth century and in the early years of the twentieth century, the so-called crisis of modern physics divided physicists and philosophers over the validity of the category of matter, which had been challenged by new discoveries about the atomic structure of matter. In the second half of the twentieth century, the scientific community was again divided over the evidence for global change until empirical data showed it unequivocally. More recently, the so-called Gaia theory has been the subject of intense debate between idealistic Gaia proponents and defenders of a materialist view of the living Earth (Kirchner, 1990; Lenton & Wilkinson, 2003). In fact, the conceptual frameworks in the natural and social sciences seem to be quite interrelated throughout history, and it cannot be otherwise, since the material reproduction of society is the ultimate determination on which ideal forms, conceptual frameworks, epistemological paradigms are based, and all of them in turn condition such material reproduction.

For Oldfield, Marxism is a useless theoretical framework for the working class because it is superfluous and demeaning. Oldfield is more comfortable with empirical descriptions of the living conditions of the working class than with theorizing, and he appreciates Engels' work on this subject. Fortunately, Engels did much more than describe the miseries of the working class in his writings on political economy and as the founder of the international labor movement. To paraphrase Oldfield, working in a factory does not qualify anyone as a Marxist, but it certainly qualifies everyone as a worker, and it may allow workers to intuit that there is some kind of labor exploitation involved, which may not be entirely clear. Since the exploitation of labor in capitalism is not immediate, as in the case of slave labor and bonded labor in earlier modes of production, where it is more direct, it is necessary to examine what are the specific forms that mediate the exploitation of labor under capitalist conditions—commodity, surplus value, wage, and so on. Hence, if workers want to overcome their labor exploitation they need to be equipped with suitable critical and theoretical tools, since a mere description of their working conditions is not enough. This is the kind of research undertaken by Marx and Marxists, and it is anything but superfluous because it aims at revealing the particular forms of labor exploitation observed on the surface of the economic phenomena and their link to the essentials of the reproduction of capital. On the other hand, in qualifying Marxism as demeaning, Oldfield seems to disregard the contribution of Marx, Engels, and many other Marxists to the international labor movement from the beginning of the nineteenth century to the present. Frankly, it is difficult to attribute decent retirement, working conditions, and many other social achievements to the generosity of capital rather than to the social pressure exerted by the working class and the labor movement.

Underlying Oldfield's critique of Marxism seems to be the usual neoliberal mantra of conflating Marxism with Stalinism. But just as we would probably not blame Jesus and the twelve apostles for the crimes of the Inquisition, neither should we blame Marx, Engels, Lenin, and Marxism as a whole for the Stalinist purges. Marxism itself is not safe from vulgarization and dogmatization, nor is Christianity or any other cultural or philosophical tradition. But while the capitalist regime consists of a society of classes based on the exploitation of labor, Marxism envisions a society without classes and without the exploitation of labor, and this makes a big difference. In general, Western society's view of Eastern socialism is a caricature of reality and deliberately distorted in terms of statistics. It is the kind of view that is taken from a sense of superiority widely consolidated in Western culture over the rest of the world, which in most cases consists of former European colonies and capitalist underdeveloped countries, assuming that the only possible progress for humanity is not only capitalist, but Western capitalist. The fact is, however, that most socialist experiences have been made in these underdeveloped countries, and the very few attempts that have been made in relatively advanced economies have been violently and rapidly crushed. For example, the Spartacist uprising in Germany and the *Consigli di fabbrica* in Italy shortly after the First World War. Such a Western supremacist view of socialist experiences can also be somewhat attributed to the so-called western Marxism (Losurdo, 2017). As a result, the Western view on Eastern socialism is usually filtered through the prism of bourgeois morality, which is assumed to be superior and universal rather than historical, a morality that bourgeois society itself has never respected and that, as Marx and Engels observed, necessarily becomes deliberate hypocrisy with the development of the productive forces:

> "The more the normal form of intercourse of society, and with it the conditions of the ruling class, develop their contradiction to the advanced productive forces, and the greater the consequent discord within the ruling class itself as well as between it and the class ruled by it, the more fictitious, of course, becomes the consciousness which originally corresponded to this form of intercourse (i.e., it ceases to be the consciousness corresponding to this form of intercourse), and the more do the old traditional ideas of these relations of intercourse, in which actual private interests, etc., etc., are expressed as universal interests, descend to the level of mere idealizing phrases, conscious illusion, deliberate hypocrisy. But the more their falsity is exposed by life, and the less meaning they have to consciousness itself, the more resolutely are they asserted, the more hypocritical, moral and holy becomes the language of this normal society". (Marx & Engels, 1975, p. 310)[18]

Oldfield credits Marxists with characterizing the consumption demands of the working class as a product of fetishism and false consciousness. However, most Marxists, at least those Marxists who can be characterized as Marxian, rarely have such a bad conception of fetishism and false consciousness. Such a misconception certainly falls into the vulgarization of Marxism and has nothing to do with either Marx's conceptualizations or those of Marxists. The level of demands of the working class is, first of all, a necessity of capital reproduction. According to this necessity, individuals are driven to consume the commodities overproduced by capital if they want to be fully integrated into capitalist society instead of remaining

---

[18] Such hipocrisy has probably reached its maximum level with the current Western support for the Palestinian genocide and the neo-Nazi regime in Ukraine.

on its margins. Second, the demand for consumption is related to the objective perception that in a capitalist society, the workers get a much smaller "slice of the pie" than the capitalists, and to the objective perception of the unequal distribution of wealth, which, conversely, is socially produced. Nevertheless, while it is true that capitalism is an inherently fetishistic and alienating system, this is in the nature of the reproduction of capital and affects all social classes, not just the working class.

4.2.2 The Concept of the Planetary Crisis in Capitalocene Narratives

The term Capitalocene was originally proposed by Andreas Malm as a substitute for the Anthropocene as the name of the geological Epoch in the Geologic Time Scale. Later on, the term underwent subsequent conceptualizations, and now it has gained some popularity in social science and the humanities. Here, the conceptual unfolding of the Capitalocene by Jason W. Moore and Donna Haraway is discussed, because these represent the characteristic epistemic views provided not only for the Capitalocene, but for other *-cenes* proposed within the humanities and social science disciplines.

According to these authors, the "Capitalocene names capitalism as a system of power, profit and re/production in the web of life. It thinks capitalism as if human relations form through the geographies of life." The Capitalocene is thus "a key conceptual and methodological move in rethinking capitalism as 'a historically situated complex of metabolisms and assemblages.'" (Moore, 2017a, p. 606, 2018, p. 240, respectively). Based on these conceptualizations of the Capitalocene, an immediate question arises: What do they add to the classical Marxist concept of capitalism as a mode of production based on the reproduction of capital through commodity production? Does this conceptualization of the Capitalocene improve anything in our scientific understanding of the general laws of social production and of capitalist production, as formulated by Marx? Does it add anything to Marx's labor theory of value, an organic theory articulated through a number of general laws accounting for the material and ideal reproduction of capitalist society? What is Marx's capital as an automatic fetish, if not an expression of the bourgeois idealism, of a system of power governing social reproduction and the alienated form of social metabolism under capitalism?

4.2.2.1 Abstraction and "Cartesian Dualism"

A recurrent issue in the Capitalocene "move" is that the Anthropocene and Earth system science approach to the planetary crisis is biased with a so-called Cartesian dualism, a "Cartesian Divide" by which humans and nature or society and nature are conceived as separate, unrelated realities that are treated as distinct analytical categories. According to the Capitalocene discourse, the Cartesian dualism attributed to Earth system science is not only methodologically and epistemologically incorrect, but modern science, in general, is based on Cartesian dualism and has become a so-called geopower that ultimately lies at the core of the planetary crisis:

> "Recognizing humans as part of nature whilst separating Humanity from Nature, troubles Anthropocene thinking at every turn. On the one hand, humans become Humanity, a singular human enterprise. They act upon, or are subject to, the "great forces of nature." On the other hand, Humanity—the upper case is deliberate—remains a geophysical force.... Practically speaking, Society is independent from Nature (Two Systems). For the earth-system scientists behind the Anthropocene, Social Factors—again, decidedly in the upper case—are added; for scholars in the humanities and social sciences, Nature is added". (Moore, 2017a, p. 597)

Rather than this dualism, the Capitalocene narrative proposes approaching the planetary crisis from a holistic-monistic perspective in which society and nature are not abstracted from each other as different categories of particular study. As Moore puts it: "The Capitalocene therefore contests social as well as environmental reductionism, and resists any periodization of capitalism derived from the mythic category of Society (humans without nature)," while "the Capitalocene argument consequently trods a different path from the governing procedures of global environmental change research: it is not a quest for 'underlying [social] causes' of environmental change, nor for connecting 'social organization' to environmental consequences". (Moore, 2017b, pp. 79 and 75)

However, Earth System science specifically asks for a "deep integration of biophysical processes and human dynamics to build a truly unified understanding of the Earth System" (Steffen et al., 2020, p. 54). Earth system science at least acknowledges that such integration is necessary, and its problem is that it has failed to achieve integration because of the incorrect theoretical and epistemological approach followed for studying human dynamics, but not for studying nature. However, the

Capitalocene discourse is unable to provide a comprehensive and integrated understanding of nature and humanity because it also navigates on wrong epistemological grounds. Substantially, the Capitalocene and other -*cenes* propose approaching the Earth System as a totality, without distinguishing between the social and the natural, humans and nature, in order to study the particularities of these realms and their mutual interactions. This is methodologically incorrect and prevents an understanding of the Earth as a dynamic system that includes human societies.

The problem "of the relation of thinking to being, the relation of the spirit to nature—the paramount question of the whole of philosophy" has had different formulations throughout the history of philosophy, including the classic body-and-soul division by Descartes, and the relationship between the material and the ideal by Ilyenkov (Engels, 1946, p. 21). This problem could not be solved by modern philosophy, until Marx and classical Marxists addressed it from a dialectical and materialist perspective. Neither the theorists of the Anthropocene nor those of the Capitalocene have been able to solve the concrete bonds between human society and nature, which, under the capitalist mode, are expressed as a planetary crisis stemming from the ontological incompatibility of humans and nature. In the case of Anthropocene theorists, this is because their approach to the natural side of the problem is based, albeit unconsciously, on dialectics and materialism, while their approach to the social side is idealistic and non-dialectical. As a result, they consider capitalism as the absolute mode of production and therefore believe there may be some compatibility between humans and nature under capitalism. Capitalocene theorists fail because they are not able to approach either side of the problem on a materialist and dialectical basis.

In the process of acquiring knowledge, humans face a material reality external to the human mind that is itself objective and independent of the way in which it is thought. Reality, no matter whether social or natural, appears to us as a concrete totality shaped by the synthesis of multiple determinations, and understanding reality is only possible through a number of mediations. First, the concrete totality has to be sensory perceived; then, it must be processed through concrete forms of thought, which are studied by logic: concepts, categories, judgments, and so on; and finally, practice determines whether the apprehension of reality by thought is correct or not. This has been long acknowledged by classic idealist philosophers such as Kant and Hegel. According to Kant's transcendental logic, the human mind formulates concepts from the sensory perception of a material reality that is diverse and contradictory. However, for Kant, diversity and contradictions are attributes of pure reason rather than of reality itself, and concepts are forms of the Kantian a priori knowledge of reality. This means that human understanding of reality is not a true historical process, and what is historical is the revelation of such a priori knowledge, but not the process of understanding itself. Hegel's dialectical logic also acknowledges the occurrence of an objective reality external to the human mind. With Hegel, contradictions as the driving force of change are explored to a level that was only preliminary in Kant. However, Hegel considered the real contradictions of nature to be bestowed by humanity's absolute spirit, and this idealist conception hampered his dialectics (Rosental, 1962).

Understanding reality is a never-ending social and historical process that, as a very general trend, proceeds from a lack of knowledge to the discovery of general and particular laws, principles, and rules governing reality. From an epistemological perspective, the process of understanding proceeds first through the perception of the studied object as a concrete totality from which some general features are extrapolated from phenomena and fixed in thought. The concrete totality is then dissected into its constitutive elements, which are abstracted from other elements in order to undertake a static analysis of their particularities; finally, the elements are reassembled to investigate their mutual concrete interactions and the history of the object. In this process, knowledge proceeds from the immediate understanding of phenomena to the conceptualization of concrete generalities conforming to the essentials of the object, its origin, and evolution. Formal logic, or intellective logic according to Hegel, plays a major role in the static analysis of the constitutive elements that are abstracted from the concrete totality. Intellective understanding is a necessary step in order to reassemble the analyzed components and investigate their interconnections and transformations. This then allows a comprehensive understanding of reality that includes the history or movement of the studied object based on dialectical logic (Rosental, 1962). Understanding the essentials of the studied object is not possible without the formal analysis of its constitutive parts, for which abstraction is necessary, or without subsequently reassembling these parts to investigate their mutual interactions and dynamics, although in the immediate process of knowledge, static and dynamic analyses are usually performed simultaneously.

Earth system science is now a discipline integrating studies from a wide spectrum of fields, such as geology, biology, chemistry, and physics. It is a discipline that is able to provide a comprehensive understanding of not only the Earth—its origins, dynamics, and possible future scenarios—but also of other planets. This includes a comprehensive understanding of the differentiation of matter on Earth, from inorganic to organic matter, and from this to "simple cells to more sophisticated single cells to multicellular life and eventually to intelligent life" (Tyrrell, 2020, p. 2). Each of the former disciplines underwent its own logical and historical evolution, abstracted from Earth's concrete totality and from other disciplines, before being integrated into an understanding of Earth's dynamics. In turn, each discipline also underwent a similar cognitive path in relation to its particular studied object. For example, botany and zoology were independent sub-disciplines of biology abstracted from one another in order to facilitate a systematic analysis of their particular subjects—an analysis that was static in regard to the wider field of life on Earth, but less static regarding their respective fields. Only later were these subfields integrated into a more dynamic view of life on Earth. However, Earth system science has not been capable until now to transcend the necessary but insufficient formal abstraction of nature and society, which would lead to a dialectical and materialist understanding of the Earth, an understanding in which the particular relationship between the capitalist Anthropos and nature—as shown by the planetary crisis—is revealed.

The Capitalocene narrative cannot provide an understanding of the concrete interrelations of society and nature that have spurred the planetary crisis, either. This is because it rejects the abstraction of nature and society as particular objects for study. However, it is not possible to conduct successful research on the modes of social organization in general, and on the capitalist mode in particular, without abstracting society from nature, just as it is not possible to develop a successful understanding of Earth's dynamics without abstracting nature from society. Only after nature and society have been independently investigated and social production laws and natural laws—that is, the general principles of social reproduction and of capitalist reproduction, and the general principles of nature and its evolution—have been understood is it possible to integrate both realms into a comprehensive understanding of the human impact on Earth and of the planetary crisis. For this reason, Marx consciously abstracted capitalist production from nature in order to investigate the universal laws of this particular mode of social reproduction, although he was completely aware of the fact that nature always has the last word in everything:

> "The use values, coat, linen, &c., *i.e.*, the bodies of commodities, are combinations of two elements—matter and labour. If we take away the useful labour expended upon them, a material substratum is always left, which is furnished by Nature without the help of man. The latter can work only as Nature does, that is by changing the form of matter. Nay more, in this work of changing the form he is constantly helped by natural forces. We see, then, that labour is not the only source of material wealth, of use-values produced by labour. As William Petty puts it, labour is its father and the earth its mother". (Marx, 1967a, p. 43)

The historical precursors of the Anthropocene designation in the GTS cannot be traced back much earlier than the eighteenth century because the planetary crisis was not a reality with noticeable phenomenological manifestations before that time. This is why neither Aristotle nor Galileo had any consciousness around a concept like the planetary crisis. Today's conceptualization of the planetary crisis is possible because it is a global reality with evident phenomenological expressions. Notwithstanding the fact that the planetary crisis started to be only an incipient reality in the nineteenth century, Marx's concepts of social metabolism and metabolic rift identified the concrete causes underlying the early manifestations of the crisis, like the rupture of the nutrient cycle and the impoverishment of soil fertility. In doing this, Marx abstracted society from nature and then reassembled the common understanding of natural sciences and natural processes available at the time with his own research on capitalist production. As a result, Marx's social metabolism and metabolic rift are concepts of his labor theory of value and are fully integrated with other concepts, such as the proletarian and capitalist classes, the industrial reserve army, profit, the concentration and centralization of capital, and so on. All of these concepts and many others are articulated *within* and *by* the general laws of capitalist production, including the law of capitalist accumulation, the law of capitalist population, and the law of the tendency of the rate of profit to fall, showing their concrete mutual determinations and concrete relation to the incipient phenomena of the planetary crisis, as revealed by the natural sciences of the nineteenth century. In summary, Marx's approach is scientific, dialectical, and materialist, and he deliberately took nature as a given to reveal the laws of capitalist production and then investigated the particular relations of capitalist production with nature by considering the early manifestations of the planetary crisis in the nineteenth century. From an epistemological and methodological perspective, it is impossible for Marx or anybody else to develop a scientific understanding of Earth, or any other real, material object, without abstractions and without proceeding through formal logic and dialectical logic.

However, the Capitalocene and other -*cene* theorists reject the operational abstraction of nature and society to understand the Earth system as it plunges into the crisis of habitability. Regarding the historical evolution of the cognition of reality and the concomitant evolution of logic, all these -*cene* approaches place themselves before Descartes and before modernity. They are closer to medieval scholasticism and various forms of mythological thought that see reality as an ideal totality in which the constitutive parts need are not distinguished through particular and formal analysis in order to carry out research on the concrete interactions among them. Because -*cene* theorists specifically deny abstracting nature and society, they are unable to undertake a critique of political economy, much less a critique based on materialism and dialectics as in the case of Marx. Economy, understood as the material production that all societies must undertake to reproduce themselves, needs to be abstracted from nature for its particular study if the concrete interactions between any socioeconomic organization and nature is to be understood. By rejecting such abstraction, and despite the references to Marx and other Marxist authors, the Capitalocene proposal lacks a proper understanding of Marx's theory of value. The Capitalocene and other -*cene* discourses are narratives with a profusion of new terms and concepts—apparently radical—that are juxtaposed without concrete determinations of causality and necessity among them. Such radical conceptualization is brought together with Marxian categories usually conceived of in an upside-down manner relative to Marx's own analysis. In this way, the Capitalocene discourse ends up with a simplistic view of the interaction between capitalist humans and nature that, paradoxically, is more reductionist than the reductionism of the Cartesian dualism that this narrative attributes to Earth system science. From an epistemological perspective, the Capitalocene and other -*cene* approach is today a form of bourgeois idealism regarding the issue of the planetary crisis.

Actually, Capitalocene and other -*cene* theorists themselves consider the Capitalocene a *narrative*, a *move*, a *discourse*, but it is a discourse filled with plenty of apparently radical neologisms that leaves untouched the fundamental problem: the material reproduction of society, that is the economy, upon which the planetary crisis is based. In this way, the Capitalocene and other -*cene* approach is closer to postmodern idealism and linguistic turns than to a materialist-based understanding of reality. Hence, it is not surprising that some of the proposals to overcome this crisis seem naïve (e.g., Haraway, 2015). It is also not surprising that the scientific understanding of nature and its fundamental laws by positive sciences—physics, chemistry, biology, geology, and so on—during modernity is seen by the Capitalocene narrative just as a "geopower," "aimed at 'discovering' and appropriating new Cheap Nature" in general, and particularly the "Cheap four" (labor, food, energy, and raw materials). In this view, the sole purpose of a scientific understanding of nature is to serve capitalist appropriation. Nature is abstracted from society and an "abstract social nature" is socially constructed, with the only function being to secure capital's access to "Cheap Natures." In other words, "abstract social nature" is an ideal creation on the basis of Cartesian dualism, and the abstraction of nature and society is aimed at allowing "geo-managerialism" of nature for the profit of capital (Moore, 2017a, 2017b, p. 610).

These unilateral and reductionist views dissolve any objective relation between society and external nature—even as a necessary abstraction—and they nullify science's ability to reflect the reality of nature in thought. This is the case (1) because nature and thought are not treated as distinct analytical categories, and (2) because science is seen as an ideal creation for the profit of capital, and not as a historically determined human universal, like labor. According to the Capitalocene discourse,

"geopower emerges at the nexus of big science, big states and "technologies of power that make territory and the biosphere accessible, legible, knowable, and utilizable." If geopower enforces Nature, it also renders Nature a motor of accumulation through the production of abstract social nature. This is accumulation by appropriation, the process of creating surplus profit via geopower and its production of abstract social nature" (Moore, 2018, p. 245).

There cannot be any doubt about the role of science in the unfolding of capitalist production on Earth, for science is a productive force in the capitalist mode, as it is in any other mode of social reproduction. Building the pyramids required knowledge of basic physics, as does designing mills to grind grain. In general, the material reproduction of society acts as the structural conditioning for the development of science at a given historical time. The reproduction of capital and competition for profit upon which the capitalist mode is based have certainly triggered the transformation of science as a productive force at an unprecedented level in human history. Labor is the specific human activity, the teleological character of which is the basis for human evolution. In the practical activity of transforming an external object, humans have to ideally prefigure the results to be obtained and the means necessary to achieve this purpose. This is the basis of a materialist understanding of knowledge, consciousness, and ethics, understood as the ability to select what is needed and what is the most suitable procedure to get the desired results. Labor is a practical activity that requires a decision-making process based on knowledge, in which knowledge and ethics reinforce one another and practice is the ultimate criterion of truth. On this basis, knowledge—including scientific knowledge—is a human universal shared by all forms of social organization.

The development of positive sciences during capitalist modernity, despite the moral reproval of some practical applications of this scientific knowledge, is an approach toward a true understanding of nature. Labor and scientific knowledge are universals that mediate human metabolism with nature, regardless of their particular historical forms. Universal forms are in a dialectical relation with historical forms, labor with abstract social labor as the substance of the value of commodities, and science with science as determined by reproduction of capital. However, dialectics requires investigating the dominant moment of dialectical unity, that is, the relationships of subsumption of the investigated pair. The reductionism of the Capitalocene narrative prevents such dialectical research and thus, any positive proposal to transcend the historical form. Undoubtedly, leaving behind the planetary crisis will require implementing a scientific understanding of Earth's dynamics, as achieved by Earth system science. The unilateral and non-dialectical view of science as solely a "geopower" and the pre-modern epistemological grounds of the Capitalocene narrative regarding the history of cognition leads to nihilism and prevents undertaking positive strategies aimed at confronting the planetary crisis.

4.2.2.2 "Accumulation by Appropriation" and Accumulation by the Capitalist Form of Labor Exploitation

Theorists of the Capitalocene and other -*cene* theorists have contributed to a better description of the "so-called primitive accumulation" explored by Marx in Capital, with research on the coercive and violent mechanisms by which non-proletarian societies are dispossessed of their means of production and subsistence, and are forced to alienate not only their labor but significant aspects of their life (Marx, 1967a, p. 713). Humans have to appropriate and elaborate upon nature through labor in order to undertake the process of social metabolism with nature. This is a human universal, regardless of the historical mode of production considered. In class societies, humans' universal appropriation of nature is based on private property. It is thus better characterized as robbery or expropriation, in which universal appropriation appears as estrangement (Foster & Clark, 2018). Thus, the expropriation of nature, goods, lands, and even humans of non-proletarian societies is a necessary moment for capital to accumulate. It is a moment that precedes the specific form of capital accumulation by labor exploitation, and that accompanies the unfolding of capital on Earth at any time, provided there is something left to proletarianize and commodify. Strictly speaking, nature is not commodified, but expropriated. What is commodified is the use value obtained *from* nature by the living activity of labor. For this to occur, nature has to first be expropriated. Once expropriated, nature may have price, but nature can only be a commodity and have value once it is processed by human labor. That is why nature is a *source* of use values, but in the absence of labor it remains just nature. Although the particular mechanisms of capitalist expropriation can be formally different, more subtle, and refined than those of former modes of production, they are not essentially dissimilar to the forms of expropriation of all class societies.

The capitalist exploitation of labor requires that labor force be a commodity, the value of which is exchanged with the equivalent value of those commodified means of subsistence necessary to reproduce labor force as such, an equivalent value that takes the form of wages. However, in capitalist expropriation there is no exchange of commodities according to the principle of equivalent values, and this is the main difference between the capitalist exploitation of labor and capitalist expropriation. Because the goal of Marx in Capital is to "lay bare the economic law of motion of modern society," the "so-called primitive accumulation" is only a secondary and mainly descriptive part of Capital, while most of the book is devoted to the specific form of capitalist accumulation through labor exploitation. As long as labor is the only activity able to produce value, the problem of capital accumulation has to be addressed from the perspective of labor, in particular, from the specific form of labor exploitation by capital. Marx is very explicit about the fact that value is only created by the living activity of labor and not by past labor contained in commodities, which enter the production process as constant capital. That is, value is not created by dead or past labor contained in constant capital, the value of which can only be activated and transferred into successive commodities by the action of living labor. Actually, Marx is very precise and, quoting Benjamin Franklin, defines the labor activity of making tools as the human concrete universal (Marx, 1967a, p. 179).

Once non-proletarian societies are alienated from their means of production, the particular form of labor exploitation that characterizes capital accumulation is the obtaining of surplus value by the two different mechanisms shown by Marx: absolute surplus labor and relative surplus labor. Surplus labor is not only specific to the capitalist mode, because in other class societies, the dominant class obtains a surplus from the labor of the dominated class. In the capitalist society, the surplus labor takes the form of surplus value, which is the *constitutive* form of capital accumulation. In fact, obtaining surplus value by labor exploitation is the process commanding expropriation, proletarianization, and commodification, and not the opposite. Hence, it is the driving mechanism for the expansion of the capitalist mode on Earth. Certainly, there have been historical periods in which the sum of capital amassed through expropriation may have eventually prevailed over the sum of capital accumulated through the specific form of capitalist labor exploitation. For example, in periods of the "so-called primitive accumulation," or, more recently, during the expropriation of public resources by capital following the collapse of Soviet-style economies. However, expropriation is only the first moment for capital to begin accumulating in non-proletarianized and non-commodified scenarios, a moment that rapidly fades once accumulation through surplus value emerges as the dominant and historical form of capital.

The question is what are the mutual determinations between these two moments, and how such determinations evolve and explain the history of the capitalist mode. For Capitalocene and some primitive accumulation theorists, the driving mechanism of capital accumulation is the so-called "accumulation by appropriation," while for Marx is the opposite: the capitalist form of labor exploitation is the driving moment for "accumulation by appropriation." The Capitalocene discourse highlights "capitalism as a history in which islands of commodity production and exchange operate within oceans of Cheap—or potentially Cheap—Natures" (Moore, 2017a, 2017b, p. 606). Although this might be formally correct, it is a rather incomplete understanding of the unfolding of capitalist production and does not account for the intrinsic forms of capital accumulation nor the history of capital accumulation.

Put simply, stating that the premise for proletarians is non-proletarians and that the premise for commodification is non-commodification is not saying much. Capital expropriates non-commodified spheres and accumulates through its form of labor exploitation once these spheres are commodified. In the history of capital there certainly may have been initially commodified "islands" in non-commodified "oceans," and the tendency of capital is to commodify everything. However, this is merely descriptive and does not grasp the essentials and causalities of the dialectical relationship between "accumulation by appropriation" (in fact, expropriation) and accumulation by capitalist labor exploitation. A dialectical and materialist understanding of capital accumulation requires knowing the concrete mutual determinations of these two processes and how these determinations evolve. This is Marx's procedure, and for this reason, Marx's conclusion regarding the dialectics of expropriation and labor exploitation is opposite to that of Capitalocene theorists.

The Capitalocene narrative conflates appropriation *without* exchange of equal value (that is, expropriation) and accumulation *with* exchange of equal value (that is, exploitation) without a proper distinction of the two processes. Therefore, it is not able to understand the causal concatenations throughout the history of capital accumulation, in which the first is a necessary moment for the second, and the second is the determinant moment over the first. As a result, the Capitalocene narrative amalgamates Marx's categories in Capital in an apparently radical discourse, but these categories are not interrelated in terms of necessity and causality within an articulated and organic theory of the capitalist production like Marx's theory of value. Thus, from the Capitalocene perspective:

"The world-historical essence of advancing labor productivity—understood in surplus value terms—is the use of Nature's unpaid work relative to labor-power.

Capitalist technology works through a simple principle: advance the rate of surplus value. The rate of surplus value turns on many qualitative and quantitative factors and conditions. But since the basic feature of rising productivity is a rising quantum of energy and raw materials (circulating capital) per unit of socially necessary labor-time, the global rate of profit depends on a threefold process: (1) material throughput must go up within the circuit of capital; (2) the necessary labor time in the average commodity must go down; (3) the costs of circulating capital (which also affect fixed capital) must be reduced (if a boom is to occur) or prevented from increasing (if a crisis is to be averted). The rate of surplus value therefore bears a close relationship to accumulation by appropriation. Accumulation crises occur when capital's demand for a rising stream of free—or low-cost—work cannot be met by human and extra-human natures" (Moore, 2018, p. 269).

Capitalism is a mode of production based on commodified interpersonal relationships spontaneously developed throughout a historical process. Its overall functioning is not subjected to any planning, and, for this reason, it has autonomous dynamics and deterministic laws similar to the laws of nature, which describe the system and its history. This is Marx's goal in Capital. Capitalist production is commodity production, and the production process is seen by capitalists through the lens of the cost–benefit equation, in which benefit or profit is the surplus value supplied by labor exploitation and cost is the capital invested to obtain profit. Every single commodity obtained from the production process contains a fraction of the total profit supplied by labor exploitation. The global profit rate is directly proportional to the total surplus value or profit and is inversely proportional to the global capital invested to obtain such profit. Capitalists only produce for profit and if there is no profit, there is no production.[19] As long as living labor is increasingly expelled from the process of capitalist production in favor of dead or past labor, which does not create surplus value, profit is obtained at a decreasing rate, which is best expressed over the long term. Hence, capitalists have to overproduce commodities not only to overcome the decreasing rate of profit but also to increase their own fraction of profit with respect to the total profit produced in society; that is, they have to compete with other capitalists for a global profit, the rate of which decreases. Essentially, these are

---

[19] "The rate of profit is the motive power of capitalist production. Things are produced only so long as they can be produced with a profit" (Marx, 1967b, p. 178).

the concrete concatenations that determine the accumulation of capital in society, its periodic crises, and the necessity of capitalist expansion on Earth. These are the fundamental mechanisms that ultimately lead to the overpopulated world, the capitalist systemic crisis, and the planetary crisis of habitability. Marx's Capital unfolds all these concrete concatenations in his labor theory of value in terms of logical categories such as causality, phenomena-essence, necessity-contingency, and through the articulation of the different capitalist laws: the law of capitalist accumulation, the law of capitalist population, and the law of the tendency of the rate of profit to fall, all of them derived from the law of value.

However, the Capitalocene view understands these concatenations upside down and/or remains ambiguous regarding the mutual determinations of categories. First, the rate of surplus value in Marx's own conception is essentially related to labor exploitation and not to "accumulation by appropriation" (that is, expropriation). In fact, Marx understands the rate of surplus value as the rate of the capitalist exploitation of labor. Second, if the "material throughput" is the physical material from nature entering the production process, stating that the rate of profit depends on a rising "material throughput" is a tautology, for it is obvious that the more capital is invested the more "human and extra-human natures" are needed. Nothing is said here about why more and more capital has to be invested and how this is related to the rate of profit, and the concrete and reciprocal relations between the rate of profit and capital accumulation remain ambiguous. From a Marxian perspective, capitalist competition for a global profit with a long-term decreasing rate requires more and more capital to be accumulated to overcome this tendency, which means more and more consumption and eventually exhaustion of "human and extra-human natures" on a planet with finite resources. Additionally, the more nature is consumed, the more waste material is returned to nature, which might end up being toxic not only due to the particular physical and chemical properties of the waste material itself but because of the quantity returned. This is Marx's metabolic rift, articulated in relation to capital accumulation and in line with the tendency of the rate of profit to fall, and thereby connected to the underlying causality and necessity embedded in the framework of Marx's theory of value.

As a matter of fact, underlying the Capitalocene narrative is a misunderstanding of capitalist crises and their relation to capital accumulation and to the rate of profit. This narrative conceives capitalist crises as driven by scarcity of natural resources: "when capital's demand for a rising stream of free—or low-cost—work cannot be met by human and extra-human natures." But it is exactly the opposite: crises are the result of an excess of the natural resources crystallized in commodities by an excess of human labor; that is, crises are the result of an excess of capital. When the overaccumulated capital reaches a threshold, it cannot be assimilated into the production processes or by individual consumption any longer. The reproduction of capital then faces a barrier that can only be surmounted by destroying the overaccumulated capital through periodic crises. This is the *law* in Marx's theory of crises, not "Marx's important but rarely discussed 'general law' of underproduction," which does not exist in Capital (Moore, 2017a, 2017b, p. 606). In other words, this is the characteristic form of capitalist crises, in which a bad harvest or lack of raw materials are contingencies affecting the general law. If the bad harvest is related to the ongoing climate change, this has to be understood within the overall dynamics of capital reproduction. Equally, if the scarcity of raw materials is related to the necessity of capital accumulation at a decreasing rate of profit or to one monopolistic strategy or another, this has to be understood within the context of capital reproduction under its deterministic laws. Obviously, when such contingencies become structural, they may become the dominant moment of capital accumulation and of crises themselves. For example, if lithium on Earth becomes nearly exhausted due to the capitalist production of batteries for solar energy, cars, etc., lithium depletion can trigger a particular crisis of capital accumulation. But it remains to be understood why and how such a lithium depletion has occurred. To claim that the depletion of nature or the difficulties of access to "Cheap Natures," be it lithium or any other natural resource, are the key determinants of capital accumulation and capitalist crises are to turn the causal concatenations and the historical unfolding of events upside down, and in this way, it becomes difficult to implement the right strategies to overcome the planetary crisis of habitability.

4.3 Overview of Current Approaches to the Habitability Crisis

Today, there is a growing number of concepts, options, and ideas, many of them based on an equally growing number of new techniques and methods, that promote solutions aimed at confronting the crisis of Earth's habitability. It is a myriad of seemingly different views of the crisis that offer innovative and particular strategies for addressing it. However, none of them touches the core of the problem, which is the immanent relation of the crisis to the capitalist mode of social reproduction and the impossibility of a long-term solution to the crisis within this mode of production. Sustainability, corporate social responsibility, recycling, circular economy, human capitalism, green capitalism, blue economy, ecological economics, degrowth, and others do not carry out an integral and comprehensive understanding of the structural relationship between the crisis and the capitalist mode. As a result, the solutions and strategies offered are only formal, and although some of the proposals can certainly alleviate the effects of the crisis, the approaches taken are usually partial and cannot change the current course of the crisis. In fact, most of these approaches seem to be a kind of Lampedusian exercise in *changing everything in order not to change anything*, which seems to be a sign of our trans-modern times.

From an epistemological point of view, the different views of the crisis are different expressions of the positivism and idealism that underlie them. They are different forms of idealism, which in turn is the bourgeois form of fetishism (Ilyenkov, 1977). First, the capitalist mode is assumed to be absolute, an ahistorical mode of social reproduction. Second, the reproduction of capital within this mode can be managed in a more humane, ecological, and sustainable way than it is now. While this may be true, the implicit assumption is that we govern the reproduction of capital and its laws, rather than being governed by it. Third, the social system can be properly managed as long as appropriate individuals, endowed with some idealized humanism and skills, are in charge. Although each individual leaves his or her particular mark on society and its institutions, regardless of whether he or she is in charge, and regardless of whether he or she is black, white, Latino, or Asian, the laws of capital reproduction must be followed, and thus the differences in management will be only formal. Fourth, the usual demand for a change in individual attitudes toward nature as a necessity for a sustainable relationship between society and nature is based on the positivist conception of society as a mere aggregation of individuals, and forgets the change in the material living conditions of society, without which the change in individual consciousness is not possible.

4.3.1 Marxism vs Other Views on the Habitability Crisis

Marxism recognizes the geo-historical determination of the habitability crisis on Earth, that is, the empirical evidence that limits this crisis to the historical period of capitalism. Moreover, Marxism has a long tradition in the critique of political economy and has conducted scientific research, both phenomenological and ontological, on the mode of production based on the sociometabolic reproduction of capital. This makes it possible to see the crisis of the Anthropocene as a result of the globalization of Marx's metabolic rift and to unravel the intimate connections between the crisis and the sociometabolic reproduction of capital. For these reasons, Marxism is in the best position to understand the structural and economic roots of the global ecological crisis, and thus to provide a theoretical and practical framework for confronting the challenge of the Anthropocene crisis.

A Marxist approach based on materialism and dialectics makes a big difference compared to other approaches such as ecomodernism and hyper-constructivist pseudo-theories like the actor network, hybrid socionatures, or Internet of Things conceptualizations, which ignore, minimize, and relativize the historical determination of the Anthropocene crisis. For example, ecomodernism promotes an optimistic, even fantastic, decoupling from negative environmental impacts through geoengineering and planet management, which will most likely lead us to a dystopian society that in many ways is already here. Hyper-constructivist pseudo-theories offer a deterministic and mechanistic view of reality in which the products of human labor have agency and are seen as actors in human life on a similar level to humans themselves. This deterministic view is expressed with a convoluted rhetoric that is far from pretending to change the world, but just the opposite, it defeats any action other than accepting the status quo (Foster, 2017). The epistemological views of the actor network and ecomodernism are precisely at the antipodes of Marxism. They operate on a crude materialist basis that turns out to be a fetishism of the current scientific and technological system, ignoring that science and technology, under the dictates of the reproduction of capital, escape human control and are far from being simply neutral. This is shown, for example, by the recent development of so-called artificial intelligence and machine learning on a business-as-usual basis by the major technological platforms.

Marxism makes also a big difference compared to degrowth. Although degrowth has correctly identified some of the causes of the Anthropocene crisis at the core of capitalist society, a theoretical deficit in understanding the fundamentals of the capitalist mode of production hinders most of the proposals of this movement, with which, incidentally, many Marxists would probably agree. The degrowth movement is closely linked to ecological economics, which constitutes the economic, supposedly scientific basis of degrowth, and which aims to challenge the dominant view of the capitalist economy from a green perspective. This movement emerged in Europe in the early 1970s, partly in response to French structuralism.[20] It recognizes the integral link between capitalism and the planetary crisis and makes the programmatic proposal that in a world that is by definition finite and with limited resources, it is necessary to reduce economic growth, which is a priori unlimited under capitalism. The main problem with degrowth and ecological economics is that both have ignored the study of political economy on a Marxian basis, that is, by unfolding the internal contradictions of capitalist production, which are invariably formally resolved in ecological economics. Marx and Marxism are usually blamed by degrowth and ecological economics for economicists and productivists, which prefer to focus on the physical and energy flows in the economy rather than on value. Because of the misunderstanding of the concept of value in ecological economics, this discipline ends up approaching the neoclassical economics it is supposed to challenge. Thus, concepts such as ecosystem services and natural capital are developed as a crude commodification and monetization of nature for the profit of capital (Costanza et al., 1997). Degrowth highlights consumption and the unequal distribution of wealth as some of the main causes of the planetary crisis. Together with exchange and production, these are moments in the process of social production that appear intertwined and determined on the surface of economic phenomena. By abstracting these moments from each other and analyzing their foundations and mutual determinations, production emerges as the main moment that gives unity and directs

---

[20] The French term "decroissance" was introduced by André Gorz. Among the main references of the "decroissance" movement and ecological economics is Nicholas Georgescu-Roegen, whose mentor was Joseph Schumpeter, a prominent economist of the Austrian school. The seminal document of degrowth is the report commissioned by the Club of Rome from researchers at the Massachusetts Institute of Technology entitled The Limits to Growth (Meadows et al., 1972).

social reproduction (Zardoya, 2009). In the absence of such an analysis, degrowth tends to focus on individual consumerism or, alternatively, to propose a list of claims that are very often based more on wishful thinking than on an objective and scientifically grounded understanding of value and capitalism as the historical culmination of the value system.[21] A steady state or degrowth economy and the decommodification of society are impossible within the limits of capital reproduction. Wealth in a capitalist society is expressed as an immense accumulation of commodities, of which money is the universal equivalent, and capital accumulates and reproduces through commodity production. Thus, neither degrowth nor decommodification is possible within the limits of capital reproduction. Although every commodity must satisfy human needs, consumerism in a capitalist society is the need for the reproduction of capital rather than the need of the people who are socially determined by this mode of production to consume the commodities produced. The only type of degrowth possible within the limits of capital reproduction is the selective one that occurs in periodic economic crises, namely the destruction of capital along with the living conditions of the working class. Therefore, decommodification and controlled degrowth are only possible outside the sociometabolic reproduction of capital. This misunderstanding has led degrowth to idealize individual subjectivity and to rely on the aggregation of individual voluntarism to confront the global ecological crisis without questioning the economic basis of this crisis. In some ways, the critique of degrowth by many Marxists is similar to Marx's critique of Proudhon for his idealism. (Iglesias, 2011)

The diagnosis of Marxism is this: the sociometabolic reproduction of capital requires the undermining of the two sources of wealth, labor and nature, an undermining driven by economic laws beyond the control of the producers. This is not something accidental in capitalism, but it is inherent in the foundations of the reproduction of capital, and this leads to a very simple conclusion: within the limits of capitalist society, it is not possible to face the Anthropocene crisis of habitability and to leave it behind for some kind of harmonious, sustainable or balanced relationship with nature. This challenge should certainly be confronted with scientific and technological measures aimed at avoiding or reducing harmful

---

[21] For examples of the misconception of value, see Kallis et al. (2013), Kallis and Swyngedow (2018). The latter is a conversation between an ecological economist and a Marxist geographer about whose Marxism many Marxists would have doubts.

environmental impacts on the Earth. However, if these measures do not take place within a socioeconomic system that has left behind, or is being driven to leave behind, the socioeconomic reproduction of capital, sooner or later they will be useless. A fundamental reason is that any technological measure must be developed within the commodity system, which is based on value and profitability. Therefore, it is determined by the economic laws of the reproduction of capital, which are the same laws that underlie the planetary crisis.

Soviet-style socialist countries and Asian countries with centrally planned economies, Yugoslavian socialism, in which cooperatives were dominant at least for a time and the state redistributed the surplus product among them, and even mixed capitalist economies in Western countries with strategic public sectors outside the circuit of capital reproduction were, to some extent and with obvious differences among them, historical experiences in which social reproduction was not entirely based on the reproduction of capital and in which society had some control over social reproduction through the state. As such, they are historical experiences that are inconceivable without the theoretical and practical contributions of Marxism to the scientific understanding of the capitalist mode of production and to the international workers movement. Therefore, many lessons can be learned from the socialist experiences on how to leave capitalism behind. However, in order to draw useful conclusions, they must be considered within the overall historical context of this mode of production and its development, and within the particular context of each experience in this scenario. Otherwise, the analysis of socialist experience becomes a vulgar exercise in throwing the baby out with the bathwater. For example, socialist experiences took place in countries where capitalism was not as developed as in the Western countries, where bourgeois revolutions had already been successful and the bourgeois class was already the ruling class of social reproduction. Moreover, socialist experiences had to face strong internal and external opposition, including economic and military aggression. Most socialist experiences occurred in underdeveloped countries, many of them former Western colonies still under pseudo-colonial regimes, such as China, Vietnam, Cuba, and others. The development of productive forces on a capitalist basis was not even an option if these countries wanted to be economically and politically sovereign from the West. In the end, no one wants to be socialist and poor, or socialist and remain a colony, which is essentially the same thing. This has been widely misunderstood by Western Marxism, which has rather often condemned these socialist experiences as productivist and non-ecological, without taking into account their particular historical context. In fact, the history of socialist revolutions in the underdeveloped world in the second half of the twentieth century was also the historical process of emancipation from Western imperialism (Losurdo, 2017). Equally important, the restoration of capitalism in socialist countries was mainly based on deliberate political decisions, so the failure of socialist experiences does not constitute empirical evidence of the failure of socialism as such. No socialist experience has been able to transcend value as the determination of the products of labor and labor power, and the capitalist-based social division of labor, including the division of manual and intellectual labor:

> "In socialism, however, we have two measures of socially necessary labor time. The main one, that of socialized labor, is time, the necessary labor time to produce a use value in given conditions.
> The other one is value, an indirect measurement of the same magnitude.
> Among those two there are continuous divergences and contradictions. An these cannot fail to emerge, for they are not due to the inability to calculate, but to the value form itself" (Iliénkov, 2012, direct translation from Spanish).

Transcending value, the commodity form, and its monetary expression are fundamental issues for the success of any socialism and communism. In summary, historical socialism was not "*developed* on its own foundations, but, on the contrary, just as it *emerges* from capitalist society; which is thus in every respect, economically, morally, and intellectually, still stamped with the birth marks of the old society from whose womb it emerges" (Marx, 1976, p. 16 emphasis in the original). It is quite likely that the seventy years of the USSR, the longest duration of any historical socialist experience, was not enough time to firmly establish the socialist foundations in society, especially when these foundations began to be slowly dismantled in the second half of the twentieth century through the introduction of market strategies and profit into the economy, until capitalism was finally restored.

Bourgeois civilization is reluctant to admit that its own creature, capital, cannot be controlled and, like all civilizations, pretends to be eternal. For this reason, and especially after the failure of socialist experiences, any attempt to transcend capitalist society is disregarded as an impossible utopia. General guidelines and some specific measures to transcend the sociometabolic reproduction of capital have already been pointed out by Marx and other Marxists (Cockshott & Nieto, 2017; Marx, 1976; Mészáros, 2000). A key point is that in a planned economy, labor time rather than money should be the unit of account for economic calculation, as this would allow for the direct exchange of individual labor without the mediation of the monetary expression of value, which would allow for the exchange of labor on a more egalitarian basis (Cockshott & Cottrell, 2007; Cockshott & Nieto, 2017; Kopanev, 2023). A nonprofit-based planned economy can direct the unfolding of productive forces toward those scientific and technological initiatives aimed at mitigating environmental degradation without the encumbrance of profit, which is something impossible in a profit-based market economy. Beyond the social division between the owners and the dispossessed of the means of production, the class structure in capitalist society is based on the fact that labor is determined as a form of value, as is the case with any commodity, and this is also the basis of the capitalist division of labor (Starosta & Fitzsimons, 2018). Transcending capitalism means transcending such a value determination. Only by transcending this value determination can labor be transformed into not just a means of satisfying needs, but satisfaction of needs, and the principle of "from each according to his ability, to each according to his needs" can be fulfilled (Marx, 1976). Such an upsurge in the material reproduction of society must be accompanied by a change in the way we approach the world: from bourgeois idealism and positivism to a materialist and dialectical worldview. It took almost 300 years of bourgeois revolutions, beginning in 1572 in Holland, for the bourgeois class to take over the political power of society and for capital to become the precondition, mediator, and result of social reproduction in many countries of the world (Callinicos, 1989). In this regard, the socialist experiences based on Marxism are relatively young, but the habitability crisis of the Anthropocene tells us to hurry up in transcending the sociometabolic reproduction of capital.

4.3.2 What Is Marxism?

In this essay, Marxism's view of the Earth's habitability crisis has been contrasted with many other views, most notably those of Earth system science and Anthropocene studies, and also with the view of Western Marxism, or parts of it, which in its various forms is the mainstream Marxism worldwide, and not without influences from postmodern thought, as this is the dominant philosophy of neoliberalism (Jameson, 1984). Therefore, the question of what Marxism is may have arisen throughout this essay.

It is certainly not an easy question to answer. There have been strong disagreements in the history of Marxism about what Marxism is and what it is not, and the history of these disagreements is beyond the scope of this section. According to Engels, Marx used to say not to be a Marxist when commenting on the supposedly materialist view of history held by French Marxists in the late 1870s, a claim that many Marxists today would probably make for themselves in response to the poor materialism of many other Marxists (Engels, 1890). Marx and Engels always defended the need for a scientific understanding of reality in order to change it according to conscious human choices based on knowledge. With the understanding of nature and society provided by the development of positive sciences in modernity, this has become a factual possibility. This, however, implies a limitation of the traditional field of knowledge of philosophy and a transcendence of the philosophy of nature and the philosophy of history as the epistemological paradigms on the basis of which nature and society were previously understood. According to Engels, with the development of the positive sciences and after Hegel's philosophical legacy, the main scope of philosophy is the study of the epistemological and logical processes by which human beings grasp the external reality outside their minds, and a philosophy of nature and of history have lost their former significance:

> "In the above, it could only be a question of giving a general sketch of the Marxist conception of history, at most with a few illustrations, as well. The proof must be derived from history itself; and, in this regard, it may be permitted to say that is has been sufficiently furnished in other writings. This conception, however, puts an end to philosophy in the realm of history, just as the dialectical conception of nature makes all natural philosophy both unnecessary and impossible. It is no longer a question anywhere of inventing interconnections from out of our brains, but of discovering them in the facts. For philosophy, which has been expelled from nature and history, there remains only the realm of pure thought, so far as it is left: the theory of the laws of the thought process itself, logic and dialectics". (Engels, 1946, p. 59)

In this respect, the distinction between a science of social history and a science of natural history is justified only on the basis of the peculiarities of both fields and on a methodological basis, but from an epistemological point of view there is only one science, the science of history:

> "We recognize only one science, the science of history. Considered from two different perspectives, history can be divided into the history of nature and the history of men. However, both are inseparable: as long as men exist, the history of nature and the history of men will be reciprocally determined".[22]

Marx realized that the historical time of only one science was yet to come and that further development of both the natural and social sciences was needed before they could become a single science of history: "History itself is a *real* part of *natural history*—of nature developing into man. Natural science will in time incorporate into itself the science of man, just as the science of man will incorporate into itself natural science: there will be *one* science" (Marx, 1959, p. 48 emphasis in the original).

Today, the planetary crisis of the Anthropocene caused by capitalist humanity makes the emergence of a "science of history" not only a possibility but a necessity if the Earth is to remain habitable for human and non-human life. A science of history, however, does not mean abandoning the various fields and subfields of science developed in modernity. On the contrary, such a science of history must be based on the positive sciences of individual fields. As in Marx's time, however, this is still a distant possibility. This is essential because a unified science of history must be based on a materialist and dialectical conception of both the social and natural realms, and the present understanding of society is still far from being materialist and dialectical. The natural sciences are forced to be dialectical and materialistic because their object of study is a material and dialectical reality. Although natural scientists are not usually aware of their own epistemological paradigm, they have to practice it unconsciously, otherwise the object of study will not be properly understood. Modern Earth system science, for example, undertakes a dialectic and materialist understanding of the Earth that accounts for the Earth's history and dynamics and allows for the reconstruction of past Earth scenarios and the prediction of

---

[22] This fragment corresponds to a draft of the German Ideology that was discarded by Marx and Engels and was not published in the 1932 MEGA edition of the German Ideology. The fragment has been directly translated by the author from the second Spanish edition of the German Ideology, published jointly by Ediciones Pueblos Unidos, Montevideo, and Ediciones Grijalbo, Barcelona, in 1974.

future ones with reasonable confidence. Certainly, the social realm is also dialectical and materialistic. However, while human action has not significantly interfered with the natural system, at least not until the habitability crisis of the Anthropocene, societies are human self-constructed structures resulting from more or less immediate human action, and this obscures a dialectical and materialist understanding of the social. Moreover, the dominant epistemological paradigm for understanding social modes of production and capitalist society is anything but dialectical and materialist, which creates additional difficulties for understanding social sciences on a dialectical and materialist basis. Therefore, while a scientific understanding of nature based on materialist and dialectical grounds is widely practiced, albeit unconsciously, there are different and contrasting epistemological approaches to society in the social sciences, and a materialist and dialectical approach is not dominant.

Marxism is the philosophical tradition, beginning in the nineteenth century, that introduces and defends a dialectical and materialist understanding of reality, whether social or natural. In other words, it is the epistemological paradigm based on the conscious materialist and dialectical understanding of the world that is necessary to change the current state of things.[23] Such an epistemology has strong implications for society—politics, economics, history, ethics—and given that the interaction of capitalist human beings with nature has evolved into a planetary crisis of habitability, it also has strong implications for nature. Marxism has focused mainly on the practical application of materialism and dialectics to the study of the social realm, for the natural realm is already studied by the natural positive sciences developed during modernity—biology, physics, geology, etc. The Marxist tradition has developed a dialectical and materialist understanding of the economy, that is, the material reproduction of society and its associated ideal forms. That is why there is Marxist economics and not, say, Marxist geology. As for the natural sciences, however, if economics is to be a scientific discipline, there is only one kind of economics regardless of traditions and schools of thought, and that is

---

[23] According to Soviet philosophers of the second half of the twentieth century, dialectics is both method, logic, and theory of knowledge, and dialectics has to be materialist or not to be (Ilyenkov 1982a, 1982b; Rosental 1962). However, this is an understanding of dialectics that is not shared by many Western Marxists, who have tended to ignore Soviet philosophy.

the science of economics. Marx's theory of value is the objective scientific theory of the commodity system based on value and is supported on dialectic and materialist grounds. This theory explains the irrationality and long-term unsustainability of the capitalist mode, which is the historical culmination of the commodity system based on value. Understanding the material and ideal reproduction of society on a scientific basis opens up the possibility of consciously shaping the material production of society, which is antagonistic to the current functioning of capitalist society. It is therefore understandable that bourgeois economics, the dominant paradigm based on epistemic grounds opposite to those of Marxism, and Marxist economics are mutually exclusive, and that the main efforts of bourgeois economics are focused on the negation of the objective character of value. It is less understandable that some Marxists do not think of the labor theory of value as a scientific theory, but this only tells of the poor dialectical and materialist understanding in the social sciences, including some Marxist schools. On the frontiers of the natural and social sciences, a dialectical and materialist understanding of psychology developed after the Second World War, especially in the former Soviet Union. For this reason, there is also a Marxist psychology that must be distinguished from the bourgeois understanding of psychological phenomena, e.g., introspectionism, behaviorism, and existentialism, that dominates in the West and is based on positivist and idealist grounds (Rubinstein, 1963).

Philosophy and epistemology have always been a major issue within Marxism. Marx's critique of classical political economy in Capital is ultimately an epistemological critique. It is a critique of the insufficient and therefore wrong approach by the classical political economy to the understanding of economic reality. Marx and Engels criticized the idealism of the young Hegelians in the German Ideology, of Proudhon in Marx's The Poverty of Philosophy, and of Dühring in Engel's Anti-Dühring. Lenin criticized the idealism of empirio-critics as Mach, Avenarius, and others in his Materialism and Empirio-criticism. Dialectical logic was developed within Marxism, mainly by philosophers of the Soviet Union but not only, as a necessity to go further in our understanding of reality given the insufficiency of formal logic. They criticized the neo-positivism developed in the West, which was essentially based on formal logic, as an epistemology that becomes functional to the capitalist system and idealizes this mode of production. Fidel Castro used to refer to the importance of ideology, his well-known battle of ideas, in transforming the bourgeois system.

REFERENCES

Bondi, D. (2015). Gaia and the Anthropocene; or, the return of teleology. *Telos, 172*, 125–137.
Bosenko, V. (1965). Por qué las ciencias naturales que conocen las formas de movimiento no pueden arreglárselas sin la filosofía. *Problemas Filosóficos De La Ciencia Natural Moderna, 3*, 62–72.
Callinicos, A. (1989). Bourgeois revolutions and historical materialism. *International Socialism, 43*, 113–171.
Chukhrov, K. (2013). Epistemological gaps between the Former Soviet East and the "Democratic West". *E-flux Journal 41.*
Cockshott, P., Cottrell, A. (2007). *Why labour time should be the basis of economic calculation.* Available at: https://users.wfu.edu/cottrell/ope/archive/0612/att-0229/01-Berlinlong.pdf
Cockshott, P., Nieto, M. (2017). *Cibercomunismo. Planificación económica, computadoras y democracia.* Editorial Trotta
Cordon, F. (1982). *La evolución conjunta de los animales y su medio.* Anthropos.
Costanza, R., d'Arge, R., De Groot, R., et al. (1997). The value of the world's ecosystem services and natural capital. *Nature, 387*, 253–260.
Costanza, R., van der Leeuw, S., Hibbard, K., et al. (2012). Developing an Integrated history and future of people on earth (IHOPE). *Current Opinion in Environmental Sustainability, 4*, 106–114.
Doolittle, W. F., & Booth, A. (2017). It's the song, not the singer: An exploration of holobiosis and evolutionary theory. *Biology & Philosophy, 32*, 5–24.
Ellis, E., Maslin, M., Boivin, N., & Bauer, A. (2016). Involve social scientists in defining the Anthropocene. *Nature, 540*, 192–193.
Engels, F. (1890). *Letter to Conrad Schmidt on 5 August 1890.* Available at: https://www.marxists.org/archive/marx/works/1890/letters/90_08_05.htm
Engels, F. (1946). *Ludwig Feuerbach and the end of classical German Philosophy.* Progress Publishers.
Engels, F. (1970). *Socialism: Utopian and scientific.* Progress Publishers.
Engels, F. (1986). *Dialectics of nature.* Progress Publishers.
Foster, J. B. (2013). Marx and the rift in the universal metabolism of nature. *Monthly Review, 65*, 1.
Foster, J. B., & Clark, B. (2018). The expropriation of nature. *Monthly Review, 69*, 1.
Foster, J. B. (2022). The return of the dialectics of nature. The struggle for freedom as necessity. *Monthly Review, 74*, 1.
Foster, J. B. (2017). Marxism in the Anthropocene: Dialectical rifts on the left. *International Critical Thought, 6*, 393–421.
Fukuyama, F. (1992). *The end of history and the last man.* The Free Press.

Haraway, D. (2015). Anthropocene, capitalocene, plantationocene, chthulucene. *Environmental Humanities, 6*, 159–165.
Hendry, D. (2004). The nobel memorial prize for clive W.J Granger. *The Scandinavian Journal of Economics, 106*, 187–213.
Iglesias, J. (2011). *La miseria del decrecimiento*. Ecologistas en acción.
Iliénkov, E. V.,., Naúmienko, L. K. (1977). *Tres siglos de inmortalidad*. Available at: https://www.marxists.org/espanol/ilienkov/tres-siglos-de-inmortalidad.pdf
Iliénkov, E. V. (2012). *La lógica económica del socialismo*. Edithor.
Ilyenkov, E. V. (1977). *Dialectical logic*. Progress Publishers.
Ilyenkov, E. V. (1982a). *Dialectics of the abstract and the concrete in Marx's capital*. Progress Publishers.
Ilyenkov, E. V. (1982b). *Leninist dialectics and the metaphysics of positivism*. New Park Publications.
Jameson, F. (1984). Postmodernism, or the cultural logic of late capitalism. *New Left Review, 146*, 62.
Jameson, F. (2003). Future city. *New Left Review, 21*, 65–79.
Janković, S., & Ćirković, M. M. (2016). Evolvability is an evolved ability: The coding concept as the arch-unit of natural selection. *Origins of Life and Evolution of Biospheres, 46*, 67–79.
Kallis, G., Gómez-Baggethun, E., & Zografos, C. (2013). To value or not to value? That is not the question. *Ecological Economics, 94*, 94–105.
Kallis, G., & Swyngedow, E. (2018). Do bees produce value? A conversation between an ecological economist and a Marxist geographer. *Capitalism Nature Socialism, 29*, 36–50.
Kirchner, J. W. (1990). The Gaia metaphor unfalsifiable. *Nature, 345*, 470.
Kirchner, J. W. (2003). The Gaia hypothesis. conjectures and refutations. *Climatic Change, 58*, 21–45.
Kopanev, G. (2023). *National automated system of computation and information processing: OGAS 2.0*. Available at: https://cibcom.org/national-automated-system-of-computation-and-information-processing-ogas-2-0/
Lenin, V. I. (1974). *Materialism and Empirio-criticism*. Progress Publishers.
Lenton, T. L. (1998). Gaia and natural selection. *Nature, 394*, 439–447.
Lenton, T. L., & Wilkinson, D. M. (2003). Developing the Gaia theory. A response to the criticisms of Kirchner and Volk. *Climatic Change, 58*, 1–12.
Losurdo, D. (2017). *Il marxismo occidentale. Come nacque, come morì, come puó rinascere*. Editori Laterza
Lukács, G. (1970). *Lenin: A study on the unity of his thought*. New Left.
Lukács, G. (1980). *The ontology of social being*. Merlin Press.
Maito, E. (2014). Piketty versus Piketty. El Capital en el siglo XXI y la tendencia descendente de la tasa de ganancia. *Revista De Economía Crítica, 18*, 250–264.

Marx, K. (1959). *Economic & philosophic manuscripts of 1844*. Progress Publishers.
Marx, K. (1967a). *Capital 1*. International Publishers.
Marx, K. (1967b). *Capital 3*. International Publishers.
Marx, K. (1976). *Critique of the Gotha programme*. Progress Publishers.
Marx, K. (1993). *A Contribution to the Critique of Political Economy*. Progress Publishers.
Marx, K., & Engels, F. (1975). *The German ideology*. Progress Publishers.
Meadows, D., Meadows, D., Randers, J., & Behrens, W., III. (1972). *The limits to growth*. Potomac Associates.
Mészáros, I. (2000). *Beyond capital: Toward a theory of transition*. Monthly Review Press.
Mooney, M., Duraiappah, A., & Larigauderie, A. (2013). Evolution of natural and social science interactions in global change research programs. *PNAS, 110*, 3665–3672.
Moore, J. W. (2017a). Anthropocenes and the Capitalocene alternative. *Azimuth, 9*, 71–79.
Moore, J. W. (2017b). The Capitalocene, part I: On the nature and origins of our ecological crisis. *The Journal of Peasant Studies, 44*, 594–630.
Moore, J. W. (2018). The Capitalocene, Part II: Accumulation by appropriation and the centrality of unpaid work/energy. *The Journal of Peasant Studies, 45*, 237–279.
Oldfield, F. (2018). A personal review of the book reviews. *The Anthropocene Review, 5*, 97–101.
Palsson, G., Szerszynski, B., Sörlin, S., et al. (2013). Reconceptualizing the 'Anthropos' in the Anthropocene: Integrating social sciences and humanities in global environmental change research. *Environmental Science & Policy, 28*, 3–13.
Piedra Arencibia, R. (2019). Marxismo y dialéctica de la naturaleza. Edithor
Piedra Arencibia, R. (2016). La dialéctica categorial y las ciencias naturales. Reseña crítica de *Proceso al azar*. *Horizontes y Raíces, 4*, 71–80.
Piedra Arencibia, R. (2018). El papel del trabajo en el desarrollo del pensamiento humano. *HYBRIS. Revista De Filosofía, 9*, 173–206.
Piketty, T. (2014). *Capital in the Twenty-First Century*. Harvard University Press.
Rosental, M., Straks, G. M. (Eds). (1960) *Categorías del materialismo dialéctico*. Editorial Grijalbo
Rosental, M. M. (1962). *Principios de lógica dialéctica*. Ediciones Pueblos Unidos
Rubinstein, S. (1963). *El ser y la conciencia Grijalbo*. Mexico
Soriano, C. (2024). The problems of the Anthropocene in the geologic time scale, and beyond. *Earth Science Reviews, 253*, 104796.

Starosta, G., & Fitzsimons, A. (2018). Rethinking the determination of the value of the labor power. *Review of Radical Political Economics, 50,* 99–115.

Steffen, W., Richardson, K., Rockström, J., et al. (2020). The emergence and evolution of Earth System Science. *Nature Reviews Earth & Environment, 1,* 54–63.

Steffen, W., Rockström, J., Richardson, K., et al. (2018). Trajectories of the Earth System in the Anthropocene. *PNAS, 115,* 8252–8259.

Thomas, J. A. (2024). The Anthropocene's stratigraphic reality and the humanities: A response to Finney and Gibbard (2023) and to Chvosek (2023). *Journal of Quaternary Science, 39,* 1–3.

Tyrrell, T. (2020). Chance played a role in determining whether earth stayed habitable. *Communications Earth and Environment, 1,* 61.

Zardoya, R. (2009). La Producción Espiritual en el sistema de producción social. In *Filosofía Marxista* (pp. 107–125). Felix Varela

CHAPTER 5

Summary and Conclusions

The ongoing planetary crisis is probably one of the greatest challenges facing humanity today. This crisis is an environmental reality that is expressed in an increasing number of parameters every day, and it is leaving an equally increasing imprint in the geological record. For this reason, geologists have considered the possibility of formalizing the planetary crisis in the Geologic Time Scale, which is the reference tool to which Earth's history is referred. However, formalizing the planetary crisis in the Geologic Time Scale is only a minor issue, no matter how important it is for geology and earth science. The big deal is that life on Earth is increasingly threatened, and this affects most living organisms, including humans. Therefore, the current crisis can be characterized as a crisis of Earth's habitability. Despite the great advances in our understanding of the Earth's dynamics thanks to the recent unfolding of Earth system science, we do not really know the full significance of the planetary habitability crisis. The inertia, feedback mechanisms, and nonlinear dynamics of the processes involved make concrete predictions somewhat speculative, and only general trends can be depicted. In this regard, caution is advised, and some technological ventures, such as ocean fertilization to capture atmospheric carbon dioxide, or weather manipulation and cloud seeding, should be viewed with caution. This is especially true when such technological measures are implemented by private companies operating

© The Author(s), under exclusive license to Springer Nature Switzerland AG 2024
C. Soriano Clemente, *Marxism and Earth's Habitability Crisis*, Marx, Engels, and Marxisms,
https://doi.org/10.1007/978-3-031-72537-1_5

under the laws of capital reproduction and with the sole purpose of profit. This can also be extended to public enterprises in many countries around the world, whose main concern is to manage the public for the benefit of the reproduction of capital.

This essay has shown that the planetary crisis is demonstrably linked to the very essence of the capitalist mode of social reproduction, a mode that consists of the reproduction of capital through commodity production. The reproduction of capital implies the unlimited accumulation of capital, and for this to happen, labor and nature must be reduced to mere things, mere commodities to be consumed in the chrematist process of capital accumulation. Consequently, the accumulation of capital on the one hand entails the degradation of nature and labor on the other, and the accumulation of capital at a decreasing rate of profit accelerates this degradation. There is no doubt about it: within this mode of production, there must necessarily be a planetary crisis, a crisis that is getting worse every day with the historical development of the mode of production. Because the reproduction of capital is governed by laws that have been formed in a historical process behind the backs of individuals, capital accumulates in society without any real human control. As a result, the social metabolism of human beings with nature is also out of control; it is an alienated form of social metabolism that corresponds to the metabolism of capital rather than that of human beings. Bourgeois thought is dominant worldwide because it is the cosmovision of the ruling class, a ruling class that is reluctant to admit that it is incapable of understanding and therefore controlling the process of social reproduction on which its dominant position is based. For this reason, the bourgeois class defends the spontaneous activity of commodified interpersonal relations as the only mode of social reproduction, and any research aimed at unfolding the contradictions of the mode of production and conceiving a planned economy is eliminated. According to bourgeois thought, the unequal distribution of wealth, recurrent economic crises, and the current planetary crisis—which are expressions, among many others, of the contradictions of the capitalist mode—are contingent features that can be managed within the limits of the mode of production, and none of them is structural to the reproduction and accumulation of capital.

The dominant bourgeois thought is deeply embedded in the current understanding of the planetary crisis, and the critical views relating the crisis to the capitalist mode are usually based more on intuition than on scientific research. With the slow dismantling of the welfare state in

Western countries and after the collapse of socialism in Europe, many of the critical views on the capitalist mode have lost their way. This includes a significant part of so-called Western Marxism, which is influenced in various ways by the postmodern philosophy of neoliberalism. Today, the dominant understanding of the planetary crisis emanating from academia, from where the supposed experts on the subject disseminate their research, is scientifically incomplete, to say the least. This is not surprising, since the truly critical views of the capitalist system have been gradually replaced in universities and research councils by formal critiques that eliminate the internal contradictions of capitalist production and leave the core of the system untouched. In this context, the emphasis on the structural determinations—the conditions and constraints, in other words, the laws—under which the material and ideal reproduction of bourgeois society take place is usually disregarded as deterministic. However, whether the dominant positivist and idealist bourgeois thought like it or not, humans are constrained by a number of laws or principles in their relationship with nature that are intrinsic to nature. In addition, the way humans understand the external reality out of their subjectivity is constrained by the laws and principles of logic, both formal and dialectical logic. Finally, humans are also constrained by the particular laws of social modes of production that humans themselves build up in their process of self-production through the different types of societies throughout history. Therefore, human beings can only try to understand the objective laws of nature, logic, and social production in order to find their best possible place in nature. This is how Marxism understands freedom: as the knowledge of necessity, of the laws inherent in reality, including the human process of thought.

The contribution of Earth system science to our understanding of the origin and evolution of the Earth is not in dispute, as is the research of the Anthropocene Working Group to formalize the planetary crisis in the Geologic Time Scale. However, the planetary crisis of the Anthropocene shows us very clearly that the current dynamics of the Earth are driven by a particular and historical human dynamic and not by an abstract humanity, and this has not been adequately addressed by Earth system science and Anthropocene studies. The empirical indicators of the planetary crisis, both environmental and stratigraphic, clearly correlate the crisis with the capitalist mode of social production. In particular, the global manifestation of the crisis in the strata roughly coincides with the golden age of capitalism worldwide after the Second World War. Despite this clear correlation, the scientific disciplines of the natural sciences, including the Anthropocene, have refused to consider serious research on the structural socio-economic roots of the planetary crisis. As a result, the understanding of the planetary crisis provided by modern geosciences is incomplete at best.

On the other hand, the social sciences' contribution to understanding the planetary crisis is influenced to varying degrees by postcolonial studies, poststructuralism, gender studies, posthumanism, and other similar subfields. At the risk of being reductionist, but for the sake of being synthetic, the social sciences' contribution to the planetary crisis is loosely influenced by postmodern thought, which is the form of bourgeois thought under neoliberalism. The net effect of postmodern thought is to dissolve any certainties about our understanding of reality. Categories such as class, human, nature, the social, the natural, mind, body, and so on are no longer valid or are minimized in a usually muddled narrative. Instead, we are supposed to operate with the totality of the observed phenomena without defining and abstracting categories for their particular analysis in order to undertake the subsequent research on their mutual interrelations. It must be said that none of the modern positive natural sciences—physics, biology, chemistry, geology, etc.—has historically proceeded this way to investigate their objects of study. If Earth System science, despite its shortcomings in understanding the socio-economic roots of the planetary crisis, is now able to provide us with a global understanding of the Earth's dynamics, it is because it has proceeded in exactly the opposite way to postmodern thought. First, it defines and abstracts categories for its particular analysis, and then it integrates the knowledge gained from a number of particular subfields with particular categories into the totality of the Earth system in such a way that the dynamics of the system and the interrelated determinations from the different fields are understood.

The main conclusion of this essay is simple: only a Marxian-based understanding of the ongoing planetary crisis can help us to delineate a mode of human social organization and a mode of human social metabolism with nature based on which we might have at least a possibility of reversing the habitability crisis on Earth. By a Marxian-based understanding it is here meant a scientific approach to the crisis based on materialist and dialectical epistemological grounds, from which the natural and social sides of the problem are not seen as simple aggregations formally interrelated, but as interpenetrating realms based on common essential grounds, on the basis of which the mutual concrete causal concatenations are established. Such an understanding, in turn, leads to another simple conclusion: only a communist organization of society, obviously reflecting the local idiosyncrasies worldwide, can drive humans into some kind of harmonious and sustainable relationship with nature. There are many arguments in support of this conclusion. Organizing the material reproduction of society on a nonprofit basis makes it possible to direct the development of productive forces toward those scientific and technological areas that aim to reduce the human impact on the earth, without the burden of profitability. Put simply, if we understand economy as the study of the material reproduction of society aimed at saving resources, a profit-based mode of production is non-economic. In a nonprofit-based production, the economic surplus absorbed by profit could be redirected to implement effective measures that mitigate environmental degradation. Given that the capitalist form of economic surplus is surplus value, the phenomenal expression of which is profit, that only labor is capable of creating surplus value, and that for this to happen labor must be a commodity bought and sold in the labor market, a communist mode of production must decommodify this fundamental human activity, on which any mode of production, not only the capitalist one, is based.

If the accumulation of capital is the ultimate cause of the planetary crisis, it cannot be the solution to the crisis, and there is no other way to accumulate capital than the capitalist way. For this reason, a communist nonprofit-based and substantially decommodified society is the only possibility—and possibility has to be emphasized here—to go beyond the current measures undertaken in the capitalist society, which aim only to mitigate the human impact on the Earth, but not to overcome the crisis of habitability and to establish a social metabolism with nature that can keep the Earth habitable. It is at least a possibility because within a communist mode of production, human beings can direct social production toward desired concrete goals. On the contrary, in the capitalist mode, social production is directed by the automatic subject of capital in favor of its accumulation, which has led us to the current planetary crisis. In his preface to *A contribution to the critique of political economy*, Marx stated:

> At a certain stage of development, the material productive forces of society come into conflict with the existing relations of production or—this merely expresses the same thing in legal terms—with the property relations within the framework of which they have operated hitherto. From forms of development of the productive forces these relations turn into their fetters. Then begins an era of social revolution.

The planetary crisis of the Anthropocene tells us that such "stage of development" is already here, and that the capitalist relations of production "fetter" the future development of humanity, actually threaten humanity itself. For this reason, the Anthropocene dilemma can be rephrased with Rosa Luxemburg's famous quote socialism or barbarism, as communism or extinction. Communism is the only theoretical corpus we have so far that is not only antagonistic to bourgeois society, but capable of positively overcoming it. It was born in the nineteenth century from within the bourgeois system, first as an ideal and later as a scientifically based potential reality. In the aforementioned preface, Marx stated: "Mankind thus inevitably sets itself only such tasks as it is able to solve since closer examination will always show that the problem itself arises only when the material conditions for its solution are already present or at least in the course of formation." This means that communism is not a utopia, as is often disregarded in bourgeois thinking, but a very real possibility, as shown by the repression of communist organizations in capitalist societies in recent history. The socialist experiences of the last century, with their differences, peculiarities, and mistakes, were attempts to overcome capitalism in the direction of a communist society. Historical socialist experiences already posed themselves transcendental questions about the life of people in a non-capitalist society, as well as a number of concrete problems regarding the material and ideal reproduction of society, too. For example, What is art? What is religion? What should they be in socialism? How do we plan the economy? How do we measure labor? In addition, there was a firm conviction in socialist thought that, despite all the horrors that accompanied the development of capitalism worldwide, it was considered overall progress in human development, and the capitalist development of productive forces, including modern sciences, was evaluated positively for any future society. Hence, the path to a communist society was not conceived as a regression to some kind of pristine past or as starting from zero, as it seems to be in some alleged alternatives to capitalism envisaged by postmodern thought, but as a transcendence of the bourgeois system while critically assimilating the best of it.

Index

A

absolute, 16, 17, 50, 56, 62, 64, 70, 78, 79, 89, 97, 100, 101, 114, 121, 122, 129, 134

abstract, 42, 43, 73, 86–90, 100, 101, 126, 127, 151

activity, 2, 10–15, 17, 28, 29, 43, 46, 50–52, 57, 72, 73, 76, 79, 87, 88, 90, 93, 94, 98, 99, 127–129, 150, 153

aggregated, aggregation, 94, 95, 97, 98, 134, 137, 153

alienation, alienated, 2, 41, 42, 44–47, 52, 79, 111, 119, 128, 129, 150

analysis, analytical, 5, 14, 17, 23, 24, 38, 40, 56, 73–75, 77, 86, 88, 96, 100, 107, 120, 122, 123, 125, 126, 137, 138, 152

antagonistic, 63, 64, 74, 75, 94, 144, 154

Anthropos, 28, 29, 31, 123

appropriation, 126, 128, 130, 131

B

bourgeois, 1–4, 43–49, 62, 66, 73, 76, 95, 98, 100, 101, 112, 113, 116, 118, 119, 125, 134, 138–140, 144, 150–152, 154, 155

Bretherton diagram, 32, 109

Bretton Woods, 60

C

Capital, 5, 39, 41, 42, 45, 53, 64–68, 70, 74, 76, 99, 100, 128, 130, 131, 133, 144

carbon dioxide (CO_2), 11, 38, 41, 149

category, categories, 5, 32, 38–40, 42, 44–47, 54, 62, 68, 70, 73, 74, 84, 85, 96, 100, 107, 116, 120, 121, 125, 126, 130, 132, 152

causal, causality, 32, 46, 54, 62, 68, 70, 72, 73, 75, 96, 106, 110, 111, 125, 130, 132, 133, 153

centralization, 61, 78, 124

circulating capital, 59, 78, 131

© The Editor(s) (if applicable) and The Author(s), under exclusive license to Springer Nature Switzerland AG 2024
C. Soriano Clemente, *Marxism and Earth's Habitability Crisis*, Marx, Engels, and Marxisms, https://doi.org/10.1007/978-3-031-72537-1

157

class, 2, 3, 16, 21, 26, 38, 58–60, 70–72, 76, 77, 102, 110, 113, 116–119, 128, 129, 137, 138, 140, 150, 152
collective, 13, 53, 88, 98
communism, 113, 139, 154
competence, competitor, 28, 48, 58, 74
complex systems, 65, 68, 69
concentration, 11, 61, 78, 124
conceptual, 24, 29, 30, 53, 54, 56, 68, 73, 76, 86, 87, 103, 104, 115, 116, 119
concrete, 29, 30, 32, 40, 42, 43, 45, 54, 56, 73–75, 86–93, 96, 99–101, 104–106, 108, 109, 112, 114, 121–125, 129, 130, 132, 149, 153, 154
conscious, consciousness, 45, 50–52, 60, 69, 78, 85, 93, 94, 108, 118, 124, 127, 134, 141, 143
constant capital, 43, 44, 48, 61, 70–72, 129
consumption, 40, 42, 58, 77, 118, 119, 132, 133, 136
contingent, contingencies, 5, 6, 72, 75, 76, 133, 150
contradiction, 3, 5, 6, 16, 26, 32, 42, 44, 46, 47, 49, 50, 53–55, 62, 63, 70, 74, 95, 98–102, 106–108, 111, 113, 118, 122, 136, 139, 150, 151
corpus, 6, 39, 54, 64, 75, 76, 84, 100, 101, 103, 108, 111, 112, 154
correlation, correlated, 14, 17, 18, 22, 25, 27, 31, 38, 39, 54, 110, 151, 152
Crawford Lake, 21, 25
culture, cultural, 20, 22, 47, 53, 57, 88, 99, 100, 117

D

dead labor, 47, 49, 57, 58, 62, 71
deduction, deductive, 23, 24, 47, 73, 77, 86, 87, 95, 96, 99–102, 104
degrowth, 134, 136, 137
density function, 68, 69
determinant, determined, determination, 14, 20, 21, 23, 24, 26, 32, 38, 40, 44, 46, 51, 52, 54, 62, 65, 68–73, 75, 78, 86, 88, 91, 92, 94, 96, 100, 101, 104–106, 108, 112, 116, 121, 124–127, 129, 130, 132, 133, 135–140, 142, 151, 152
deterministic, 38, 91, 93, 104, 111, 131, 133, 135, 151
diachronic, diachronism, diachronous, 24, 25
dialectical, 4, 24, 25, 40–42, 46, 62, 66–68, 73, 74, 84, 86, 92, 98–101, 105–108, 114, 116, 121–125, 127, 130, 140–144, 151, 153
dichotomy, dichotomic, 24, 31, 39, 40
dilemma, 84, 85, 105, 111, 114, 154
discipline, 5, 7, 8, 14, 20, 22, 23, 27, 37, 84, 89, 96, 103, 105, 111, 112, 114, 119, 123, 136, 143, 152
dual, dualist, dualistic, 4, 6, 8, 18, 32, 42, 103, 105, 114
dynamics, 7, 10, 11, 26, 31, 46, 50–52, 54, 58, 60, 68, 69, 75, 76, 78, 87, 90–93, 106, 109, 111, 113, 120–123, 127, 131, 133, 142, 149, 151, 152

E

ecological, 20, 38, 39, 41, 112, 134–137
ecomodernism, 135

economy, economic, 5, 13, 19, 31, 32, 39, 42, 44–49, 52, 55–57, 60–63, 65, 66, 68, 69, 72–76, 78, 95–98, 100, 101, 112, 116, 117, 125, 126, 128, 134–140, 143, 144, 150, 153, 154
empiricism, empirical, 2, 4, 9, 12–14, 24, 27, 29, 37–39, 47, 49, 61, 65, 69–71, 73, 78, 85, 86, 88, 90, 94, 95, 97, 98, 100–102, 104, 106, 110, 113, 114, 116, 135, 139, 151
environment, environmental, 11, 12, 14, 16, 20, 23, 25–27, 32, 37–39, 41, 77, 78, 109, 120, 135, 138, 140, 149, 151, 153
epistemic, 64, 98, 99, 102, 105, 112, 119, 144
essence, essential, 13, 20, 25, 26, 30–32, 40, 42, 43, 45, 47, 50, 54, 57, 62, 68, 73, 74, 78, 83, 87, 91, 97, 98, 110, 111, 115–117, 122, 130–132, 138, 142, 144, 150, 153
ethics, 57, 79, 92, 93, 127, 143
etymology, etymological, 27, 28
evolution, 2, 5, 12, 13, 16, 17, 23, 30, 38, 40, 46, 50, 53, 54, 57, 66, 67, 69, 85, 86, 88–92, 98, 106–108, 122, 123, 125, 127, 151
exchange, 12, 13, 42–44, 51, 52, 57, 74, 78, 92, 128, 130, 136, 140
exegesis, 67, 68
expropriation, 128–130, 132
extinction, 1, 2, 10, 11, 38, 154

F

fallacy, fallacious, 95, 97, 102, 109
fascism, 3, 112
fetishism, fetishist, 41, 44–46, 98, 118, 134, 135

financial capital, sphere, sector, 46, 59–62
formal, formally, 6, 21, 24, 30, 32, 57, 62, 65, 73, 74, 89–91, 95–98, 101, 105, 122, 123, 125, 128, 130, 134, 136, 144, 151, 153
freedom, 151

G

geoengineering, 135
geosphere, 8, 53
Global boundary Stratotype Section and Point (GSSP), 17–19, 21, 24, 25, 28–30
global change, 8, 61, 116
Great Acceleration, 12, 29, 31, 38
Great Ordovician Biodiversification, 25
Great Oxidation, 25
greenhouse, 11, 17, 41

H

history, 2, 3, 5, 7–20, 22, 24–27, 30, 31, 37, 38, 40, 44, 47, 49–54, 57, 58, 65, 66, 71, 77, 79, 84–86, 88, 90, 91, 94, 96, 97, 103, 106, 108, 109, 114, 116, 121, 122, 127, 129–131, 139, 141–143, 149, 151, 154
Holocene, 11, 12, 19, 21, 28, 31
human, humanity, 1, 2, 4–15, 17, 21–30, 31, 37–39, 44–47, 50–53, 55, 57, 58, 62, 71, 76–79, 83–85, 87, 88, 90–94, 97–99, 102–109, 111, 114, 117, 119–123, 126–129, 131–135, 137, 141–143, 149–154

I

ideal, 17, 42, 44, 45, 51, 53, 57, 58, 88, 93, 94, 99, 100, 112, 116, 119, 121, 125, 126, 143, 144, 151, 154
image, 55, 76
immanent, 5, 24, 50, 52, 75, 78, 88, 92, 96, 98, 110, 112, 134
impact, 8, 10–14, 16, 25, 26, 29, 37–39, 103, 109, 123, 135, 138, 153
indeterminacy, 68–70
individual, 2, 12, 20–22, 44, 47–49, 51, 52, 58, 60, 69, 77, 85, 88, 93, 94, 96–99, 106, 118, 133, 134, 137, 140, 142, 150
induction, inductive, 23, 24, 73, 86, 87, 90, 94–96, 100
industry, industrial, 11, 15, 52, 74, 77, 78, 124
inorganic, 53, 54, 92, 106, 107, 123
International Commission on Stratigraphy (ICS), 8, 18, 19, 21, 27, 28
International Union of Geological Sciences (IUGS), 8, 18, 19, 21, 28
intersection, 95

J

judgment, 86, 95, 121

L

language, 57, 95, 99–101, 106, 107, 118
life, 1, 2, 27, 28, 38, 42, 45, 46, 51, 52, 62, 63, 65, 66, 78, 89, 90, 94, 106, 118, 119, 123, 128, 135, 142
living labor, 47, 49, 50, 57, 58, 62, 70, 71, 129, 131

logic, logical, 24, 27, 47, 54, 62, 72–74, 77, 86, 87, 95, 99–102, 104, 107, 108, 121–123, 125, 132, 141, 143, 144, 151

M

magnitude, 2, 9, 10, 12, 15, 26, 37, 39, 43, 53, 57, 62, 89, 91, 139
market, 23, 42, 60, 62, 96, 97, 139, 140, 153
Marx-Engels Gesamtausgabe (MEGA), 64
mass extinction, 2, 10, 17, 28, 53
material, 3, 4, 21, 39, 42, 44, 51, 53, 57, 58, 60, 66, 70, 76, 77, 87, 90, 92–94, 96, 98, 100, 105, 106, 108, 113, 116, 119, 121, 122, 124–126, 131, 132, 134, 140, 142–144, 151, 153, 154
matter, 2, 5, 12, 23, 25, 38, 41, 51, 52, 69, 84, 85, 91–93, 106, 107, 116, 121, 123, 124, 132, 149
mechanistic, 41, 68, 69, 135
mediation, 44–47, 51, 52, 58, 140
metaphysics, metaphysical, 72, 84, 94, 106
methane (CH_4), 11
method, methodological, 5, 17, 27, 40, 56, 66, 73–75, 86, 105, 114, 119–121, 125, 133, 142
money, 42, 44, 46, 48, 59–62, 69, 70, 75, 100, 137, 140
monist, monistic, 4, 84, 85, 92, 120
Montreal Protocol, 41
moral, morality, 3, 14, 92, 94, 113, 118, 127, 139

N

necessity, 5, 54, 56, 62, 68, 71, 72, 75, 76, 96, 103, 110, 118, 125, 130, 132–134, 142, 144, 151

neoliberalism, neoliberal, 59, 60, 63, 117, 141, 151, 152
Neolithic, 24
Nobel laureate, 7, 47
non-dialectical, 62, 115, 121, 127
notion, 94–96, 99, 101, 114

O

objective, objectively, 72, 83, 87, 91, 98, 99, 105, 106, 110, 119, 121, 122, 126, 137, 144, 151
ontology, ontological, 4, 5, 39, 42, 96, 121, 135
organic composition, 48–50, 70, 77
organic theory, 40, 48, 72, 86, 119, 130
overaccumulation, 58, 61, 77
overpopulation, 40, 52, 71
overproduction, 58, 76, 77

P

paradigm, 4, 40, 64, 85, 93, 102, 105, 112, 116, 141–144
paradox, paradoxical, paradoxically, 72, 79, 87, 110, 113, 114, 125
phenomena, phenomenon, phenomenal, phenomenological, 25–27, 31, 32, 42, 45, 47–49, 54, 56, 57, 61, 62, 64, 67–69, 72–76, 85–87, 90, 92, 94–98, 100, 101, 104, 117, 122, 124, 132, 135, 136, 144, 152, 153
plate tectonics, 54, 67, 75, 90, 91
plutonium, 15, 21
political economy, 47, 66, 67, 73, 74, 100, 101, 108, 116, 125, 135, 136, 144
population, 10, 12–14, 23, 41, 51, 52, 55, 66, 68, 69, 71, 124, 132
postmodern, postmodernism, 40, 88, 126, 140, 151, 152, 155

power, 21, 42, 45, 46, 52, 57, 72, 79, 119, 120, 126, 131, 139, 140
power law, 69, 70
practice, practical, 2, 4, 6, 8, 17, 18, 22, 23, 27, 42, 46, 50, 51, 53, 54, 57, 60, 64, 76, 79, 83, 86–88, 90–93, 95, 99, 100, 105, 107, 111, 112, 120, 121, 127, 135, 138, 142, 143
prediction, predicted, 23, 24, 86, 108, 142, 149
price, 42, 46, 48, 55, 60, 68, 74, 96, 97, 128
principle, 5, 26, 54, 60, 66, 71–73, 76, 77, 86, 87, 91, 93, 95, 96, 107, 122, 123, 128, 131, 140, 151
probabilistic, 69
productive capital, sphere, sector, 59–62
productive forces, 21, 25, 47, 72, 79, 118, 126, 127, 138, 140, 153, 154
productivity, 43, 47, 49, 50, 52, 59, 69, 71, 72, 75, 130, 131
proletarian, 124, 130

Q

qualitative, 15, 25, 26, 53, 71, 72, 131
quantitative, 12, 15, 25, 26, 29, 53, 68, 71, 72, 131
Quaternary, 20, 22, 28

R

radionuclides, 21, 25
rate, 2, 10–12, 15, 32, 44, 47–50, 55, 56, 58–65, 67–72, 74–78, 102, 108, 124, 131–133, 150
reality, 1, 4, 21, 24, 26, 46, 48, 54, 55, 65, 73, 76–78, 83, 84,

86–88, 90, 93–96, 98, 99, 101, 105–108, 116, 117, 121, 122, 124–126, 135, 141–144, 149, 151, 152, 154
reasoning, 24, 70, 73, 86, 87, 95
reductionism, reductionist, 120, 125–127, 152
reflection, reflect, 4, 14, 19, 20, 27, 65, 74, 76, 84, 103, 106, 107, 126
Russian, 19, 20

S

scientometrics, 95, 96
Second World War, 21, 49, 54, 99, 144, 152
sensory perception, 76, 90, 94, 122
social being, 12, 13, 50, 53, 71, 88, 97–99, 105
socialism, socialist, 84, 113, 117, 118, 138–140, 151, 154
society, 2–4, 20, 21, 24, 31, 38, 39, 42–46, 48, 50, 55, 58, 60, 66, 76, 77, 83, 84, 88, 92–96, 98–100, 102, 103, 112–114, 116–121, 123–126, 128, 129, 131, 132, 134–144, 150, 151, 153, 154
Soviet, 20, 99, 113, 129, 138
species, 1, 2, 10–13, 38, 39, 50, 51, 69, 89, 106
spontaneous, spontaneously, 3, 60, 69, 72, 93, 96, 97, 108, 131, 150
stewardship, 109, 111, 112
stochastic, 68
strata, 2, 14–16, 18, 23–31, 91, 152
subjective, subjectively, 22, 72, 93, 98
subsistence, 13, 42, 50, 51, 59, 72, 128

surplus value, 43, 44, 46, 48–50, 54, 57–59, 62, 69–72, 100, 117, 129–132, 153
sustainable, sustainability, 4, 56, 134, 137, 153

T

tautology, 97, 132
technical composition, 70
teleological, 24, 51, 52, 57, 79, 88, 91, 93, 127
temperature, 11, 38, 90
theoretical, 5, 6, 24, 38–40, 48, 64–68, 72–74, 76, 83, 84, 86, 87, 89–91, 95, 99–101, 103, 104, 106, 115–117, 120, 135, 136, 138, 154
thought, 42, 48, 57, 65, 76, 86–88, 92, 95, 99, 101, 105–108, 113, 114, 121, 122, 125, 126, 141, 143, 150–152, 154, 155
totality, 40, 86, 95, 96, 121–123, 125, 152
toxicity, 77, 78

U

unilateral, 126, 127
Union of Soviet Socialist Republics (USSR), 20, 139
United States (US), 19, 20
universal, 39, 40, 46, 51–53, 57, 58, 71, 74, 79, 88, 93, 101, 118, 123, 126–129, 137

V

valorization, valorized, 3, 43, 46, 48, 58, 59, 63, 68, 78, 100, 101
value, 5, 6, 39–49, 52, 54–63, 65–68, 70, 72–75, 77, 78, 84, 90, 92, 97, 98, 100–102, 111, 119, 124,

125, 127–130, 132, 136–140, 144
variable capital, 43, 48, 60, 78
varved sediments, 21
verbal, 57, 64–66, 87, 94, 99, 104

W

wage, 45, 49, 50, 52, 60, 73, 117, 128
war, 3, 4, 19
wealth, 43, 52, 98, 102, 119, 124, 136, 137, 150
West, Western, 2–4, 19, 46, 99, 113, 117, 118, 138, 139, 144, 151
work, 41, 57–59, 65, 75, 93, 110, 116, 124, 130, 131, 133
world, 1–4, 9, 10, 12–14, 19, 20, 38, 41, 42, 44, 45, 51, 53, 55, 57, 58, 60, 69, 71, 77, 83, 92, 110, 111, 117, 132, 135, 136, 139, 140, 143, 150